Advances in Materials for Electrochemical Energy Applications 2024

Advances in Materials for Electrochemical Energy Applications 2024

Guest Editor

Claudio Mele

MDPI

Basel • Beijing • Wuhan • Barcelona • Belgrade • Novi Sad • Cluj • Manchester

Guest Editor
Claudio Mele
Department of Engineering
for Innovation
University of Salento
Lecce
Italy

Editorial Office
MDPI AG
Grosspeteranlage 5
4052 Basel, Switzerland

This is a reprint of the Special Issue, published open access by the journal *Energies* (ISSN 1996-1073), freely accessible at: https://www.mdpi.com/journal/energies/special_issues/2D8B2DU2KK.

For citation purposes, cite each article independently as indicated on the article page online and as indicated below:

Lastname, A.A.; Lastname, B.B. Article Title. *Journal Name* **Year**, *Volume Number*, Page Range.

ISBN 978-3-7258-7376-0 (Hbk)
ISBN 978-3-7258-7377-7 (PDF)
https://doi.org/10.3390/books978-3-7258-7377-7

Contents

About the Editor

Claudio Mele

Claudio Mele is an Associate Professor of Applied Physical Chemistry at the Department of Engineering for Innovation of the University of Salento. He completed his degree and Ph.D. in Materials Engineering at the University of Lecce. He is the head of the Applied Electrochemistry Laboratory at the Department of Engineering for Innovation of the University of Salento.

His research activity mainly focuses on the application of electrochemical and spectroelectrochemical techniques in the research, development, and characterization of materials for electrochemical energy (batteries, fuel cells, supercapacitors); the research, development, and electrochemical characterization of metallic biomaterials; the study of the corrosion of metallic materials and alloys for applications in the petrochemical, biomedical, electronics, industrial, and cultural heritage sectors; the research, development and characterization of anticorrosive coatings; and the study of the electrodeposition of metallic materials and alloys.

Preface

This Reprint, "Advances in Materials for Electrochemical Energy Applications", reflects the necessity of bridging fundamental materials research with scalable solutions, from metal–air batteries to hydrogen production, ammonia synthesis, and photovoltaic energy capture. The works also highlight the need for stable, efficient, environmentally sustainable, and economically viable materials.

Claudio Mele
Guest Editor

energies

Editorial

Advances in Materials for Electrochemical Energy Applications

Claudio Mele

Department of Engineering for Innovation, University of Salento, Via Monteroni, 73100 Lecce, Italy; claudio.mele@unisalento.it

Sustainable and environmentally friendly energy storage and conversion technologies are essential to satisfy the dramatically increasing global energy demand and to reduce dependence on non-renewable fossil fuels [1,2]. The development of novel materials plays a key role in improving the properties and performance of devices in a wide range of electrochemical energy applications, including batteries, supercapacitors, flow batteries, fuel cells, hydrogen storage, photocatalysis, and thermal energy storage [3,4]. Electrochemical energy technologies continue to play a defining role in the transition to a low-carbon society [5–7]. The intricate mechanisms at electrochemical interfaces, including charge transfer and surface reactions, have been explored with advanced spectroscopic techniques that probe interfacial phenomena with high specificity and sensitivity [8–10]. In particular, surface-specific nonlinear optical methods have been demonstrated to reveal adsorption and reaction processes at electrified surfaces, offering insights applicable to catalyst and electrode design for energy conversion systems [11–16]. Electrochemical characterization techniques such as cyclic voltammetry and impedance spectroscopy further complement these insights by providing detailed information on charge transfer kinetics and interfacial phenomena [17,18]. The Special Issue "Advances in Materials for Electrochemical Energy Applications" gathers nine papers reflecting the necessity of bridging fundamental materials research with scalable solutions, from metal–air batteries to hydrogen production, ammonia synthesis, and photovoltaic energy capture, as shown schematically in Figure 1. These papers also highlight the need for stable, efficient, environmentally sustainable, and economically viable materials.

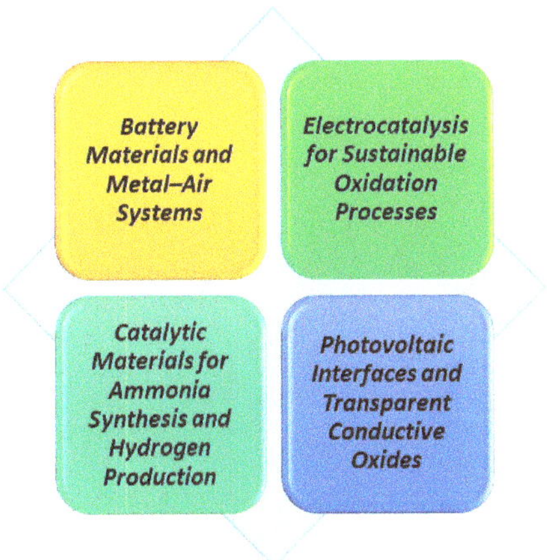

Figure 1. Paper topics presented in this Special Issue.

1. *Advances in Battery Materials and Metal–Air Systems*

Three contributions address innovation in battery technologies, particularly metal-based and metal–air systems, which are promising for their high energy density and abundance of active materials.

Bagheri et al. [19] presents a study on zinc electrodes in zinc–air batteries proposing a protective coating strategy for neutral electrolyte environments, offering significant improvement in long-term stability and electrochemical performance. The work addresses a number of challenges in zinc–air batteries, such as anode corrosion, zinc dendrite formation, and electrolyte carbonation [20]. These issues are tackled by developing a photopolymerizable organic–inorganic hybrid coating for zinc electrodes and using a neutral NaCl electrolyte. This innovative coating combines the advantages of organic and inorganic components, offering enhanced corrosion resistance, hydrophobicity, and durability. The findings demonstrate how targeted surface engineering may overcome persistent challenges such as dendrite formation and electrode degradation, improving the stability of zinc electrodes and enhancing their rechargeability.

Metal–air systems are further represented by the study of Zhuk et al. [21] on the development of magnesium–air (Mg–air) batteries with increased power using cost-effective commercially available magnesium alloy anodes. The paper focuses specifically on the MA2-2 alloy (AZ31 type), exploring the use of a concentrated 15 wt.% NaCl electrolyte to reduce internal resistance and improve battery performance. The approach offers a safer, affordable alternative to Al–air and Zn–air batteries for emergency and backup power applications. The innovative aspect lies in its adapting engineering solutions from Al–air batteries to Mg–air systems and leveraging commercially available materials for cost-effectiveness and environmental safety. While Mg–air batteries are less energy-efficient than their Al–air counterparts, their safety, availability, and lower cost make them competitive for emergency and backup power applications [22]. The study reveals notable improvements in power output, suggesting a straightforward pathway toward performance enhancement through readily available materials.

Progress in magnesium-based energy storage is highlighted through the investigation of Arbenin et al. [23] on Prussian Blue analogues (PBAs) based on 3d metals as cathode materials for magnesium-ion batteries. The work provides a compelling link between crystal structure, ion transport, and electrochemical behavior, positioning multivalent battery systems as a realistic future alternative to lithium. The study highlights the influence of the properties of 3d metals on the composition, structure, and electrochemical performance of these materials. It addresses the key challenges of magnesium-ion batteries, including the low mobility of Mg^{2+} ions due to strong ionic interactions, structural stability during ion intercalation/deintercalation, and low specific capacity. The study establishes clear correlations between structural properties and electrochemical performance, contributing to the development of efficient and environmentally friendly magnesium-ion battery technologies.

2. *Electrocatalysis for Sustainable Oxidation Processes*

Three articles focus on electrocatalytic oxidation, targeting key substrates such as glycerol, urea, and ethanol, molecules of broad industrial relevance and potential value in electrochemical conversion systems. These contributions reflect a shared scientific direction: the drive to develop catalysts that are more active, durable, cost-effective, and compatible with green chemistry principles [24].

The paper of Nascimento Silva et al. [25] focuses on the development of core–shell M@Pt/C (M=Co, Ni, Sn) electrocatalysts for efficient glycerol oxidation in alkaline media, aiming to enhance the performance of Direct Glycerol Fuel Cells (DGFCs). The study

https://doi.org/10.3390/10.3390/en19041008

introduces an innovative synthesis method using sodium borohydride reduction to create nanostructured catalysts with a 1:1 atomic ratio of metal to platinum. The core–shell structure enhances catalytic performance by modifying the electronic properties of the platinum shell and creating more active sites, improving activity and stability. It also reduces platinum usage, lowers costs, and increases resistance to poisoning by intermediates. The obtained results highlight the potential of core–shell structures for designing highly active and stable electrocatalysts for clean energy applications.

Further progress is demonstrated through a one-pot electrochemical synthesis of ternary Ni-Cu-Fe particles, designed for efficient urea oxidation. The method emphasizes simplicity, rapid processing, and strong catalytic output. The paper of Wala-Kapica et al. [26] focuses on the development of a novel ternary Ni-Cu-Fe electrocatalyst synthesized through a fast and simple one-pot electrodeposition method for improved urea oxidation. The study addresses the need for efficient and cost-effective anodic materials to advance the commercialization of DUFCs, which are promising for sustainable energy production. The innovative aspect lies in the incorporation of iron into the Ni-Cu system, which enhances the electrocatalytic activity and stability of the material. The synthesized Ni-Cu-Fe powder demonstrated a unique coral-like dendritic morphology, providing a high active surface area. Electrochemical tests revealed that the material exhibits significant activity toward urea oxidation, with a stable potential of 2.38 V vs. RHE for 3 h at a current density of 10 mA cm^{-2}. While the activity is lower compared to some advanced catalysts, the simplicity and scalability of the synthesis method make it a promising approach for future development of efficient electrocatalytic materials. The Ni-Cu-Fe catalyst shows improved activity over Ni-Cu-based materials but lower performance than advanced catalysts, with the advantage of being a simple and scalable synthesis method.

Wala-Kapica et al. [27] propose a complementary investigation on sustainable ethanol oxidation, highlighting the significant influence of graphene oxide in NiCuGO composite catalysts. By integrating carbon-based structures, their paper underscores the potential of hybrid materials to improve electrocatalytic efficiency. The authors focus on designing sustainable ethanol oxidation catalysts using a NiCuGO composite material, which incorporates nickel, copper, and graphene oxide. The study highlights the innovative use of graphene oxide in the electrodeposition process to enhance the catalytic activity and surface area of the material. The research demonstrates that the NiCuGO20 composite, prepared with a higher current density, exhibits the highest peak current density for ethanol oxidation (8.6 mA cm^{-2} at 0.66 V vs. Ag/AgCl). This performance surpasses many similar nickel-based nanomaterials, showcasing the potential of the proposed catalyst for efficient and eco-friendly ethanol fuel cells. The method is also notable for its rapid and environmentally friendly synthesis process, which excludes toxic substances and allows for precise control over material properties.

3. *Catalytic Materials for Ammonia Synthesis and Hydrogen Production*

Two papers extend the theme of chemical energy conversion into industrially relevant reactions such as ammonia synthesis and hydrogen generation. Both contributions demonstrate that electrochemical and thermochemical transformations increasingly converge toward sustainability-driven design strategies.

The study of Zybert et al. [28] examines the stability of a highly active cobalt-based catalyst for ammonia synthesis, focusing on performance degradation and structural evolution. The work provides new knowledge for improving catalyst lifetime in high-temperature, high-pressure environments. The paper focuses on the stability studies of a highly active cobalt catalyst (CoCeBa) for ammonia synthesis, a process critical for hydrogen storage and energy applications. The innovative aspect lies in the use of a cobalt-based catalyst promoted with cerium and barium, which demonstrates superior activity

https://doi.org/10.3390/10.3390/en19041008

compared to traditional iron catalysts. The study employs an accelerated aging method to simulate industrial conditions, revealing that the CoCeBa catalyst is highly resistant to high temperatures and maintains stable performance over long-term operation. Key findings include less than 10% activity decline after 436 h of overheating at 600 °C, no significant changes in surface area, morphology, or element distribution, and the ability to operate effectively at temperatures between 400 and 470 °C. Additionally, the method of Ba promoter addition does not significantly impact the catalyst's properties, allowing for flexibility in production methods. This research highlights the potential of CoCeBa catalysts for industrial ammonia synthesis with improved efficiency and stability.

The work of Mezulis et al. [29] investigates the kinetics of water–aluminum scrap reactions for green hydrogen production, using a two-step approach to optimize yield and efficiency. This research stands out for its integration of materials recycling with renewable fuel generation. The paper focuses on studying the reaction kinetics of aluminum hydrolysis using aluminum scrap and alkali solutions (up to 1 M) to optimize green hydrogen production on small and medium scales. The innovative aspect lies in the use of flat aluminum waste chips, which allow for a simpler planar one-dimensional shrinking core model to describe reaction kinetics, as opposed to more complex spherical models. Key findings include achieving a hydrogen yield of 1145 mL per 1 g of aluminum with 1 M NaOH, which is 92% of the theoretical maximum. The study also highlights the economic impact of alkali concentration and grade, emphasizing the importance of minimizing alkali loss for cost-effective hydrogen production.

4. *Photovoltaic Interfaces and Transparent Conductive Oxides*

This Special Issue also includes the comprehensive review of Musztyfaga-Staszuk et al. [30], discussing fabrication methods for front contacts to transparent conductive oxide layers in photovoltaic structures. The paper focuses on reviewing methods for producing front contacts to Transparent Conductive Oxide (TCO) layers in photovoltaic structures, emphasizing Atomic Layer Deposition (ALD) and Chemical Vapor Deposition (CVD) techniques. TCO materials like ZnO, SnO_2, and In_2O_3 are highlighted for their unique properties, such as high electrical conductivity and transparency, which are crucial for improving solar cell efficiency. The study emphasizes the importance of minimizing electrical and optical losses in TCO layers and explores innovative approaches like laser micro-machining to enhance the quality of metallic components and reduce resistance. Key findings include predictions for future advancements in solar cell technology, such as increased efficiency, reduced production costs, and the prevalence of Passivated Emitter and Rear Cells (PERCs) in mass production by the year 2040. The research also identifies ALD as a promising method for depositing homogeneous, ultra-thin TCO layers with controlled thickness and low surface roughness, despite its time-consuming nature.

The article critically evaluates technological pathways, device performance implications, and fabrication trade-offs. This work enriches the Special Issue by providing a broad industrial and scientific perspective on interface engineering, extending the scope to solar-energy generation.

5. *Conclusions and Outlook*

The nine publications included in this Special Issue reflect a dynamic and rapidly advancing research landscape and provide concrete guidelines for the development of more efficient, cost-effective, and environmentally friendly energy technologies. From metal–air batteries and multivalent ion storage to electrocatalytic oxidation and hydrogen production, the collected works highlight the central role of materials science in shaping the future of energy technologies. Key research directions emerging from this Issue include the following:

https://doi.org/10.3390/10.3390/en19041008

- Advances in electrode stability and surface modification;
- Expansion of multivalent battery chemistries;
- Development of catalytic materials for selective oxidation and energy conversion;
- Integration of recycled resources into hydrogen production;
- Continued optimization of photovoltaic interface engineering.

Looking forward, the field will benefit greatly from interdisciplinary approaches that incorporate materials design, electrochemistry, advanced modeling, and scalable manufacturing. The editor expresses his gratitude to the authors and reviewers for their contributions to this Special Issue and hopes that this collection will inspire further research into electrochemical materials and sustainable energy solutions.

Funding: This research received no external funding.

Acknowledgments: The Guest Editor would like to acknowledge Assistant Editor of *Energies*, for his continued support, ensuring the success of this Special Issue. The staff of the *Energies* Editorial Office are also acknowledged for their collaboration, together with all the contributors and reviewers.

Conflicts of Interest: The author declares no conflicts of interest.

References

1. Karol, V.; Sharma, P.; Singh, A.; Goel, D.; Kaur, S. Review of Energy Storage Devices: Fuel Cells, Hydrogen Storage, Rechargeable Batteries, PV Solar Cells. In *Materials for Boosting Energy Storage. Volume 3: Advances in Sustainable Energy Technologies*; ACS: Washington, DC, USA, 2024; Chapter 1, pp. 1–15.
2. Enasel, E.; Dumitrascu, G. Storage solutions for renewable energy: A review. *Energy Nexus* **2025**, *17*, 100391. [CrossRef]
3. Lobato-Peralta, D.R.; Okoye, U.P.; Alegre, C. A review on carbon materials for electrochemical energy storage applications: State of the art, implementation, and synergy with metallic compounds for supercapacitor and batteries electrodes. *J. Power Sources* **2025**, *617*, 235140. [CrossRef]
4. Sun, S.; Sun, X.; Chen, Z.; Liu, Y.; Wilkinson, D.P.; Zhang, J. *Carbon Nanomaterials for Electrochemical Energy Technologies: Fundamentals and Applications*; CRC Press: Boca Raton, FL, USA, 2017.
5. Dunn, B.; Kamath, H.; Tarascon, J.M. Electrical energy storage for the grid: A battery of choices. *Science* **2011**, *334*, 928–935. [CrossRef] [PubMed]
6. Tarascon, J.M.; Armand, M. Issues and challenges facing rechargeable lithium batteries. *Nature* **2001**, *414*, 359–367. [CrossRef]
7. Liu, J.; Zhang, J.G.; Yang, Z.; Lemmon, J.P.; Imhoff, C.; Graff, G.L.; Li, L.; Hu, J.; Wang, C.; Xiao, J.; et al. Materials science and materials chemistry for large scale electrochemical energy storage: From Transportation to Electrical Grid. *Chem. Rev. Adv. Funct. Mater.* **2013**, *23*, 929–946. [CrossRef]
8. Li, C.Y.; Tian, Z.Q. Sixty years of electrochemical optical spectroscopy: A retrospective. *Chem. Soc. Rev.* **2024**, *53*, 3579–3605. [CrossRef]
9. Bozzini, B.; Amati, M.; Gregoratti, L.; Mele, C.; Abyaneh, M.K.; Prasciolu, M.; Kiskinova, M. In-situ photoelectron microspectroscopy during the operation of a single-chamber SOFC. *Electrochem. Commun.* **2012**, *24*, 104–107. [CrossRef]
10. Magnussen, O.M.; Groß, A. Toward an atomic-scale understanding of electrochemical interface structure and dynamics. *J. Am. Chem. Soc.* **2019**, *141*, 4777–4790. [CrossRef]
11. Baldelli, S. Probing Electric Fields at the Ionic Liquid−Electrode Interface Using Sum Frequency Generation Spectroscopy and Electrochemistry. *J. Phys. Chem. B* **2005**, *109*, 13049–13051. [CrossRef]
12. Bozzini, B.; D'Urzo, L.; Mele, C.; Busson, B.; Humbert, C.; Tadjeddine, A. Doubly Resonant Sum Frequency Generation Spectroscopy of Adsorbates at an Electrochemical Interface. *J. Phys. Chem. C* **2008**, *112*, 11791–11795. [CrossRef]
13. Bard, A.J.; Faulkner, L.R. *Electrochemical Methods: Fundamentals and Applications*; Wiley: Hoboken, NJ, USA, 2001.
14. Bozzini, B.; D'Urzo, L.; Mele, C. Electrodeposition of Cu from cyanoalkaline solutions in the presence of CPC and PEG. An electrochemical and in Situ SERS investigation. *J. Electrochem. Soc.* **2005**, *152*, C255–C264. [CrossRef]
15. Shen, Y.R.; Ostroverkhov, V. Sum-frequency vibrational spectroscopy on water interfaces: Polar orientation of water molecules. *Chem. Rev.* **2006**, *106*, 1140–1154. [CrossRef]
16. Bozzini, B.; Gianoncelli, A.; Kaulich, B.; Mele, C.; Prasciolu, M.; Kiskinova, M. Electrodeposition of manganese oxide from eutectic urea/choline chloride ionic liquid: An in situ study based on soft X-ray spectromicroscopy and visible reflectivity. *J. Power Sources* **2012**, *211*, 71–76. [CrossRef]
17. Trasatti, S.; Petrii, O.A. Real surface area measurements in electrochemistry. *Pure Appl. Chem.* **1991**, *63*, 711–734. [CrossRef]

https://doi.org/10.3390/10.3390/en19041008

18. Huerta, F.; Mele, C.; Bozzini, B.; Morallon, E. Voltammetric and in Situ FTIRS Study on CN^{-} and $Au(CN)^{-}_{x}$ Complexes at the Polycrystalline Gold Surface in Citrate Medium. *J. Electroanal. Chem.* **2024**, *569*, 53–60. [CrossRef]
19. Bagheri, S.; Bozzini, B.; Esposito Corcione, C.; Striani, R.; Mele, C. Protective Coating for Zinc Electrodes of Zinc-Air Battery in a Neutral Electrolyte. *Energies* **2025**, *18*, 5599. [CrossRef]
20. Mainar, A.R.; Colmenares, L.C.; Grande, H.J.; Blázquez, J.A. Enhancing the Cycle Life of a Zinc–Air Battery by Means of Electrolyte Additives and Zinc Surface Protection. *Batteries* **2018**, *4*, 46. [CrossRef]
21. Zhuk, A.; Belyaev, G.; Borodina, T.; Kiseleva, E.; Shkolnikov, E.; Tuganov, V.; Valiano, G.; Zakharov, V. Magnesium-Air Battery with Increased Power Using Commercial Alloy Anodes. *Energies* **2024**, *17*, 400. [CrossRef]
22. Zhang, T.; Tao, Z.; Chen, J. Magnesium–air batteries: From principles to applications. *Mater. Horiz.* **2014**, *1*, 196. [CrossRef]
23. Arbenin, A.; Egorov, S.; Prikhodko, I.; Fedorova, A.; Penkova, A.; Selyutin, A. Prussian Blue Analogues Based on 3d-Metals as Cathode Materials for Magnesium Ion Batteries. *Energies* **2025**, *18*, 711. [CrossRef]
24. Braunchweigh, B.; Hibbits, D.; Neurock, M.; Wieckowski, A. Electrocatalysis: A direct alcohol fuel cells and surface science perspective. *Catal. Today* **2013**, *202*, 197–209. [CrossRef]
25. Silva, R.N.; Pocrifka, L.A.; Antolini, E.; Ribeiro Passos, R.R. Electro-Oxidation of Glycerol on Core-Shell M@Pt/C (M = Co, Ni, Sn) Catalysts in Alkaline Medium. *Energies* **2025**, *18*, 305. [CrossRef]
26. Wala-Kapica, M.; Gąsior, A.; Maciej, A.; Smykała, S.; Kazek-Kęsik, A.; Baghayeri, M.; Simka, W. One-Pot Fast Electrochemical Synthesis of Ternary Ni–Cu–Fe Particles for Improved Urea Oxidation. *Energies* **2024**, *17*, 5455. [CrossRef]
27. Wala-Kapica, M.; Szewczyk, M.; Simka, W. Designing Sustainable Ethanol Oxidation Catalysts: The Role of Graphene Oxide in NiCuGO Composite Material. *Energies* **2024**, *17*, 288. [CrossRef]
28. Zybert, M.; Ronduda, H.; Patkowski, W.; Rybińska, V.; Ostrowski, A.; Sobczak, K.; Raróg-Pilecka, W. Stability Studies of Highly Active Cobalt Catalyst for the Ammonia Synthesis Process. *Energies* **2023**, *16*, 7787. [CrossRef]
29. Mezulis, A.; Richter, C.; Lesnicenoks, P.; Knoks, A.; Varnagiris, S.; Urbonavicius, M.; Milcius, D.; Kleperis, J. Studies on Water-Aluminum Scrap Reaction Kinetics in Two Steps and the Efficiency of Green Hydrogen Production. *Energies* **2023**, *16*, 5554. [CrossRef]
30. Musztyfaga-Staszuk, M.; Czupryński, A.; Radev, R. Review of the Chosen Methods of Producing Front Contacts to Transparent Conductive Oxides Layers in Photovoltaic Structures. *Energies* **2022**, *15*, 9026. [CrossRef]

https://doi.org/10.3390/10.3390/en19041008

energies

Review

Review of the Chosen Methods of Producing Front Contacts to Transparent Conductive Oxides Layers in Photovoltaic Structures

Małgorzata Musztyfaga-Staszuk [1,*], Artur Czupryński [1] and Rossen Radev [2]

[1] Welding Department, Silesian University of Technology, Konarskiego 18A, 44-100 Gliwice, Poland
[2] Department of Materials Science and Technology, Faculty of Mechanical and Manufacturing Engineering, University of Ruse, 8 Studentska St., 7017 Ruse, Bulgaria
* Correspondence: malgorzata.musztyfaga@polsl.pl

Abstract: It is well known that PV thin films can be deposited by an extensive range of more or less expensive and complicated techniques (such as sputtering, chemical vapor deposition (CVD), physical vapor deposition (PVD), pulsed laser deposition, atomic layer deposition (ALD)). The present paper focuses on TCO layers applied by chosen techniques, including mainly the ALD and CVD methods. Thin layers of transparent conductive oxides constitute a well-known group of materials with unique properties. Oxides such as ZnO, SnO_2, and In_2O_3 are the most significant materials of this type; some of them are discussed in the paper. From the application point of view in the photovoltaic industry, the goal is to apply a method that will provide the highest value of electric charge conductivity while maintaining the minimum value of absorption in the layer and a reduced value of the reflection coefficient. The implementation of significant achievements in the coming decade is for developing guidelines for metallization processes and TCO layers deposited by the ALD method. The work contains chosen engineering processes, including the fabrication of transparent conductive oxides (TCO) thin films applied to silicon substrates by ALD and CVD for application as emitter conductive coatings in photovoltaic structures and the fabrication front metallization of solar cell using different techniques, including among others laser techniques. Moreover, the work also contains predictions about solar cells, which will be among the most prevalent solar cells in mass production using thin- and thick-film technology.

Keywords: transparent conductive oxides (TCO); atomic layer deposition (ALD); solar cells

1. Introduction

Progress in producing photovoltaic cells with ever-higher efficiency has been made in recent years, mainly thanks to modern techniques and materials, enabling the intensive development of individual stages of their production [1]. Innovative, which cannot be seen at first glance, able to effectively convert sunlight into electricity, turn out to be transparent conductive layers—transparent to light and at the same time perfectly conductive to electricity [2]; the photovoltaic sector (PV) has enormous development potential, and can be considered a new and extremely promising branch of the global industry. The most significant factor in the growth of the photovoltaic market is the constant reduction of the cost of obtaining 1 W_p of power from the energy converter. This means that producers of PV panels/modules must achieve higher energy conversion efficiency parameters, namely increasing the conversion efficiency of the solar cell and the photovoltaic (PV) module, as well as reducing costs in the process of production [3].

The sun is an inexhaustible energy source, so the current challenge that scientists worldwide are facing is to increase the efficiency of cells—maximize the value of the quotient: cell efficiency/production cost. The development of photovoltaics is the answer to the growing demand of the energy sector for new renewable electricity solutions, primarily due to the energy transition in Poland resulting from the global pandemic and the war with Ukraine. However, this development will not be possible without systemic solutions, i.e., financial support from the state and local authorities and research work in this area (Table 1). Renewable sources are slowly beginning to play an increasingly important role in this development, but also gas sources and the PV photovoltaic sector has enormous development potential. The article's authors focused on one application criterion in technology development in the chosen area of the photovoltaic industry, item 3 of Table 1.

Table 1. Technology development in various areas of the photovoltaic industry [4].

No	Application Criteria	Specific Tasks to Be Performed
1	Absorber	Development of performed research on contamination and technical defects of modules/panels Development of thinner and larger wafer areas for more efficient use Development of new passivation techniques
2	Photovoltaic cells	Developing lower-cost manufacturing solar cells Designing high-precision equipment for accurate solar cell processing Improving anti-reflection layers and light-trapping in the solar cells
3	Metal grids (contacts)	Reducing surface recombination by minimizing contact surfaces and strong doping under contacts
4	Interconnection	Developing modern ways of joining solar cells
5	Packaging	Innovative drive to reduce optical losses Developing changes in the appearance of modules Improving inspection systems to detect defects, automatically record their identified class, and reduce them while maintaining high sensitivity, efficiency, and process performance Developing new materials to reduce the cost of creating modules while maintaining their operational reliability
6	Manufacturing	Ongoing innovation in commercially used products Introducing changes at every stage of production to improve the finished product

Production technologies and the application of surface layers have a stable position and are regarded as basic knowledge in the field of material engineering. Many research centers worldwide working on surface engineering are applying all possible means to describe the phenomena taking place on the surface of solids. Such a situation instigates scientists to undertake challenging research to determine the effect of conditions deposition of TCOs layers by various techniques and minimize electrical and optical losses occurring in functional layers of the entire ready electronic structure.

In light of the analysis of the literature review, the authors set out two new directions worthy of attention for further research. As new, experimentally obtained results in the temperature field of applying transparent conductive oxides on the contact metallization, one of the current research directions is the deposition of emitter conductive coatings, so layers of Transparent Conductive Oxides (TCOs) in photovoltaic structures by atomic layer deposition (ALD) [5].

On the other hand, the innovative nature of the research will concern the development of process engineering, which can include the presented dependence of the influence of lasermicro-machining on the improvement of the quality of the metallic component by minimizing the resistance of the connection of this front component with the substrate. Laser techniques are used in many areas of industry, including, among others, cutting, drilling, drawing, inspection and detection of quality defects in photo-elements, and shaping of material properties and structure [6]. Laser technology in producing photovoltaic cells with high efficiency in converting solar radiation into electricity is becoming an indispensable component of modern photovoltaic technology [7]. The components and functional layers of a silicon photovoltaic cell can include the layer forming the emitter of the solar cell, the passivation and anti-reflection layer, and the front and back electrodes [8,9].

According to the requirement of the EU Directive 2009/28/EC of the European Parliament and the Council, the increase in the use of renewable energy sources (RES) in the balance of final energy will be up to 20% in 2030. So the direction of the measures taken under this article is in line with the implementation of the Polish Energy Policy until 2030 [10].

2. Metal Grids

One of the current research topics in scientific units is the reduction of covering the front side of the cell by electrode contact paths, minimizing the so-called cover factor. In 2016, the width of the printed path of the front electrode was lower than 50 μm, while the same in-line printing was reduced to a level below 20 μm [11]. In 2031, it is expected that the width of the path will be reduced from 34 to 20 μm, and overprint in line with greater precision will be reduced from 10 to 6 μm. In 2021, the number of busbars in these solar cells was from 5 to 6, but in 2031 it will increase to more than 12 [11]. Figure 1 presents the market share of single- and double-sided solar cells. It is expected that from 2025 to 2025, the share of double-sided solar cells will prevail, although the share of single-sided solar cells will remain at about 40%.

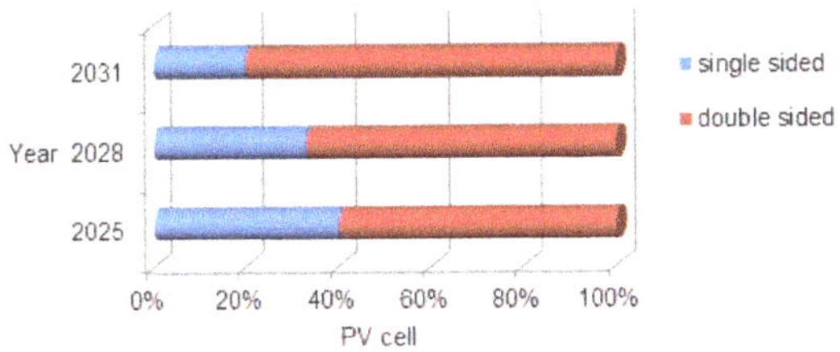

Figure 1. The market dominance of single-double-sided contact concepts [3].

Figure 2 shows some estimations concerning reducing silver pastes/inks, the most expensive materials used in c-Si cell technologies. It is predicted during the following years to continue the reduction of silver per PV cell. In 2021 it was 80 mg for standard Passivated Emitter and Rear Contact (PERC), both single- and double-sided PV cells in the mentioned format in Figure 2. In 2031 we expected a reduction down to 50 mg per PV cell [3].

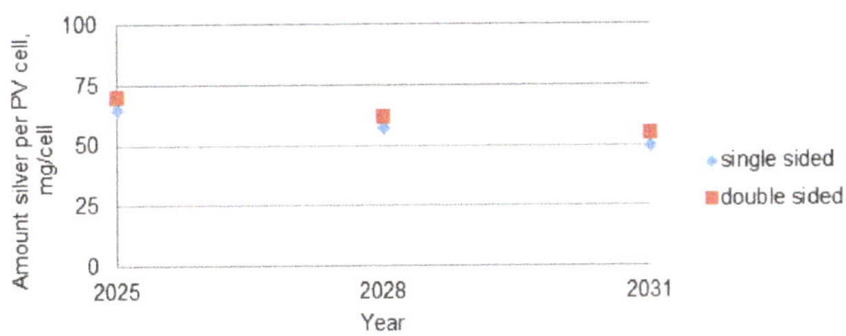

Figure 2. The trend for remaining silver per PV cell for chosen concept p-type (front and rear side of PV cell, where value for cell size are 166 and 166 mm^2) [3].

The electrode application process must meet the requirements of, among others, industrial, such as high efficiency and low cost of production. It does not include photolithography, vacuum deposition techniques, and many chemical methods, as they involve too many manufacturing steps [12]. In practice, various industrial techniques can be applied to the front electrode [13]. The most common of these is the screen printing method, which is quick and straightforward and uses pastes based on silver and aluminium; it has been used since 1975 [14]. Since then, this method has evolved rapidly and has become the dominant technique in the photovoltaic industry. Yet the coating process of silicon layer using silk-screen printing is time-consuming and challenging to automate. The height adjustment of the coated layer necessitates the application of additional screens, which translates itself into additional costs. The mechanical load applied on the wafer/insert during electrode coating can bring about numerous cracks in it and other damage. Saving on the material through the recovery of its surplus as compared to other methods is impossible in this method since only pastes can be applied here instead of, for example, powders. Taking into account material factors and fabrication conditions (having an impact on quality, in particular with the application of pastes and powders) as well as economic factors, it seems to be grounded that further experimental research involving the initial phases of the process should be carried out [15–17].

Today, 95% of solar cells are metaled using this method. Compared to other methods, it is anticipated that the front electrode screening technique will remain the leading technique. In recent years, the design of screen printers has been improved, making it possible to produce more than 1500 cells per hour [18]. Lack of process repeatability with smaller path widths can significantly reduce the efficiency of solar cells by the influence of factors such as paste viscosity and clogging of the mesh or leaving traces of the screen mesh. It is assumed that forming metal contacts using the stencil printing technique is one of the most cost-effective solutions for producing high-efficiency solar cells [19].

In the case of a template, printing is easy. A template with a given shape is placed directly on the selected substrate, and, for example, the printed paste is pressed through the holes in the pattern with a squeegee (fabric, piece of wood). After removing the template, a printed pattern remains on the surface, representing the template [9,20]. We can be found in the literature the following printing methods: double print, which is known as print on print; the single print, which is understood as one screen print process for the whole metallization, as well dual print process, which is printed separately in two different printing processes both busbars and paths. Figure 3 shows predictions for using these methods for the deposit of silver pastes in the coming years. Although the share using the single print method will decrease, it will continue to have a significant market share. In comparison, the double print method will have a small impact on the market share of no more than 15%.

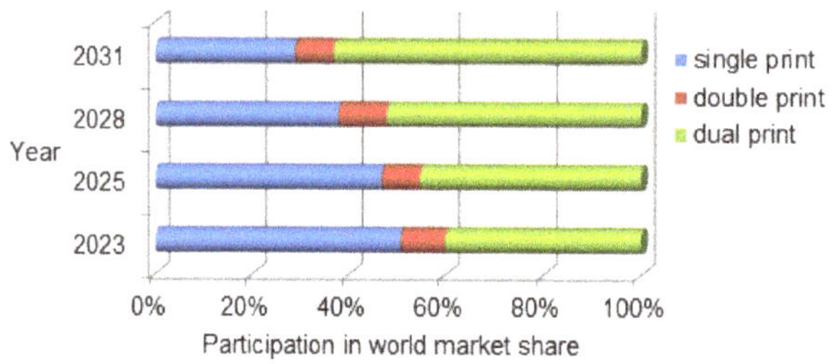

Figure 3. Prediction of the use of various front contact printing methods [3].

Compared to screen printing, new laser technology is quickly becoming one of the most promising and best alternative track application techniques. Laser technology in producing solar cells is becoming a necessary element of the modern, essential part of an automated operation in producing photovoltaic cells. Some examples involve the fabrication of front metallization using laser methods presented in available literature [16,21–23]. Solar cell technology is achieved using an advanced approach like selective emitter structures with the laser beam LDSE (Laser-doped Selective Emitter) [24]. More advanced solar cell concepts include [25] metal or emitter wrap-through (MWT/EWT) [26,27], laser-fired contacts (LFC) [28] or interdigitated back contact (IBC) [29]. We can also find in the literature works on the optimization of metallization grids (for instance, the geometry) [30,31]. Of particular importance is the proper selection of the material of the electrode and substrate. Based on the data in [32], it is possible to find precisely 56 companies involved in the product of metallization paste, one of the essential supply items for PV cell producers. Based on Figure 4, we find information that the most significant number of companies producing silver paste (more than 60%) and the smallest number of copper paste (less than 8%).

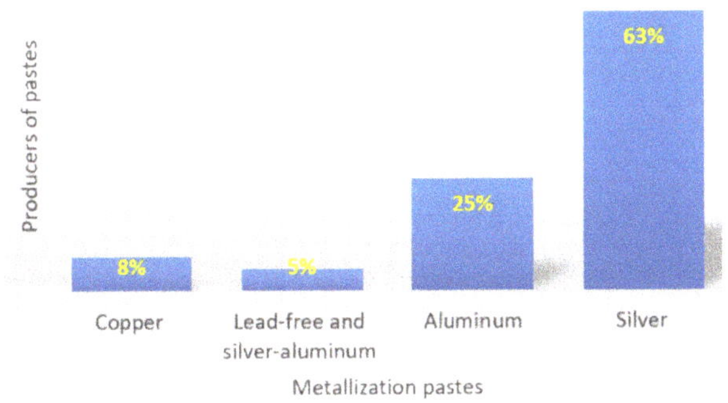

Figure 4. Market survey on metallization pastes [32].

Currently, the photovoltaic market, specifically the part of the market responsible for pastes, is dynamically developing to help the customer and thus better adapt the offer to their expectations and requirements. Improving the efficiency of solar cells requires the application of low-temperature pastes in hetero-junction technology, manufacturing contact with emitters of high sheet resistance, reducing the consumption of paste applied for front and the rear connecting contacts, and applying advanced printing technologies (for example, dual printing is increasingly being implemented into production) [33].

3. Application of ALD and PVD Methods

The second of the current research directions is the deposition of conductive emitter coatings in photovoltaic structures using the ALD method, a variant of the CVD method [5]. The ALD process relies on the sequential release of the precursors in gaseous or liquid form to deposit the coating layer by layer. The first precursor is introduced into the reaction chamber, where it then forms a monolayer on the surface of the substrate. The chamber is then purged with an inert carrier gas (usually nitrogen or argon) to remove unreacted precursors and reaction by-products, and another precursor is introduced into the chamber to form another layer, followed by purging again [34]. The described phenomena are successive stages of one ALD cycle (Figure 5), and the entire coating process consists of many such cycles.

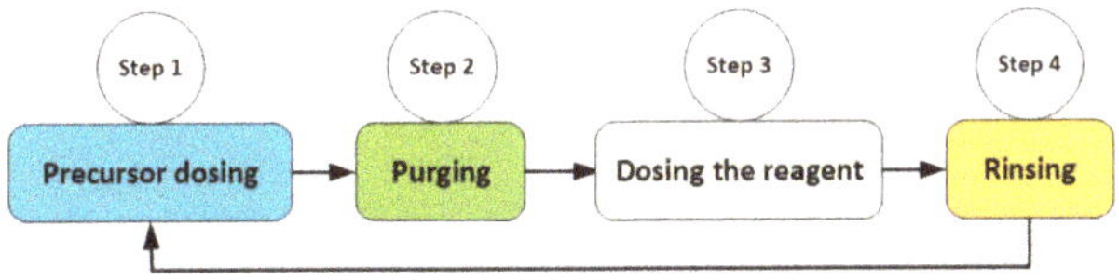

Figure 5. ALD cycle diagram.

Among the characteristic features of the ALD method are the sequentially of the process and the self-limiting rate of layer growth. The former is based on the fact that the chemical precursors are introduced alternately into the growth chamber, and after each introduction of the reactant, the chamber is purged with an inert gas. Consequently, the combination of precursors occurs only at the surface of the substrate, unlike in the classical CVD method, where the reactants are introduced simultaneously, and the reactions between them occur in the gas phase. Thus, the ALD technique's layers are more homogeneous [35].

The coating application process consists of many cycles [34–38]. The ALD method can be used to deposit [36]:

- Single elements, e.g., from group IV of the periodic table,
- Binary compounds, e.g., metal oxides,
- Multi -component compounds, e.g., hydroxyapatite.

The precursors in the ALD process should have the following characteristics [39]:

- Volatility —gas pressure >0.1 Tr at <200 °C, liquid at evaporation temperature without decomposition,
- Reactivity —they can react quickly with the substrate and the other precursor in a self-limiting manner (most are air-sensitive),
- Stability -the precursor must retain physical and chemical properties at high temperatures,
- By -products-which should not react with the coating and be easy to remove from the reactor chamber,
- Availability at the lowest possible price.

In this process, it is easy to control the growth of the applied layer by controlling technological parameters such as, among others: substrate temperature, time of introducing precursor into the chamber, purging time after each reagent, the precursor temperature, and the number of cycles. The temperature is the most critical parameter that allows controlling the surface coverage directly. ALD processes occur at low temperatures below 350 °C [34,39,40]. The temperature range in which a saturation effect occurs with layer growth (one monolayer is formed in one cycle) depends on the ALD process in question and is referred to as the "ALD growth window". The temperature outside the "temperature window" generally results in poor growth rates and deposition of non-ALD layers due to low kinetics of the reactions taking place or too much condensation of the precursor on the substrate (at low temperature), or due to thermal decomposition or rapid desorption of the precursor (at high temperature). Schematically, the "growth window" is shown in Figure 6. To achieve the maximum benefits offered by the ALD method, it is crucial to realize the process within the limits of the characteristic "temperature window" [34,36].

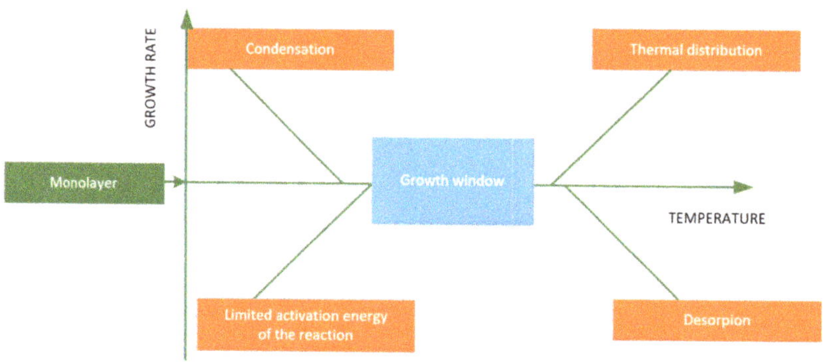

Figure 6. Temperature dependence of layer growth rate [36].

The rate of chemical reactions in the ALD process, and thus the film growth rate, is strongly temperature-dependent and can vary in nature. Diagrams of the possible types of dependencies in the ALD window are shown in Figure 7 [40].

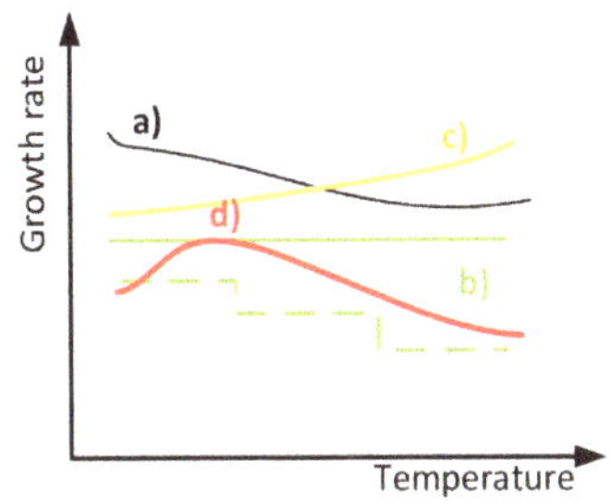

Figure 7. Dependence of growth rate on the temperature in the ALD window: (**a**) with increasing temperature, the growth rate decreases; (**b**) the growth rate is constant; (**c**) with increasing temperature, the growth rate increases; (**d**) the growth rate decreases with increasing temperature after reaching a maximum value [40].

Deposition of ALD thin films is carried out in specially adapted reactors. There are four different types of reactors [39]:

- With a closed chamber system,
- With an open chamber system,
- With a semi-closed chamber system,
- With a semi-open chamber system.

Currently, the most common type is a closed-chamber reactor. Layers are deposited in a growth chamber (reaction chamber), into which reactants are alternately introduced. Initially, they can be in the solid, liquid, or gas phase. The carrier gas in the process is an inert gas in whose atmosphere the deposition of layers takes place [39].

It can be stated, comparing them to the conventional coating application methods, that they has many advantages (Figure 8), including the possibility of using highly reactive precursors, homogeneity of the obtained layers and control of their thickness, and low surface roughness [39,41,42].

ALD	CVD
Ability to use highly reactive precursors	Use of less reactive precursors
Precursors interact with substrate separately	Precursors interact with the substrate at the same time
Precursors cannot decompose at process temperature	Precursors can decompose at process temperature
The homogeneity of the obtained layers is ensured by the saturation mechanism	The homogeneity of the layers depends on the reactant flow and temperature
The thickness of the coating depends on the number of cycles of the process	Precise control and monitoring of process parameters allows control of coating thicknesses
Excessive dosage of the precursor is allowed	The amount of precursor dosage is very important

Figure 8. Comparison of ALD and CVD processes [39–41].

The main disadvantage of the ALD process is the limited range of materials that cannot be highly reactive and the selection of precursors characterized by low activation energy to prevent the thermo-degradation of the substrates [41]. The ALD process is time-consuming because the coating is created from many ultra-thin layers. The required thickness of the coating is obtained by repeating ALD cycles many times (even tens of thousands of times). The typical growth rate of the achieved layer is 100 to 300 nm per hour. Using specially designed reactors, it is possible to increase the deposition rate to 1–2 μm/h. On the macroscopic scale, the process can be related to laminated coatings; however, the ALD process does not use adherents [41]. The coating materials used in the ALD process are extensive (Table 2), and the obtained layers may be [39]: single-element, two-component compounds, or multi-component compounds. Using ALD technology, it is possible to obtain mono-crystalline, polycrystalline, or amorphous layers.

Table 2. Materials used to make ALD thin films [34,40,43].

Pure Elements	Oxides	Nitrides	Sulfides	Carbides	Fluoride
C, Al, Si, Ti, Mn, Fe, Co, Ni, Cu, Zn, Ga, Ge, Mo, Ru, Rh, Pd, Ag, Sb, Ta, W, Os, Ir, Pt	Al_2O_3, TiO_2, Ta_2O_5, Nb_2O_5, Zr O_2, HfO_2, SiO_2, SnO_2, In_2O_2, ZnO, MgO, La_2O_3, Y_2O_3, CeO_2, Sc_2O_3, Cr_2O_3, Er_2O_3, VO2, B_2O_3, Co_2O_3, CuO, Fe_2O_3, NiO, Ga_2O_3, WO_3, …	AlN, TaNx, NbN, TiN, MoN, ZrN, HfN, GaN, WxN, InN, …	ZnS, SrS, CaS, PbS, …	TiC, NbC, TaC, …	CaF_2, SrF_2, ZnF_2, …

4. Use of TCO in a Photovoltaic Structure

The third of the current research directions is related to transparent semiconductor oxides. For several years, this has been an exciting area of interest for many scientists and research centers [44]. These materials are of particular interest to the field of optoelectronics and photovoltaics [45]. This is confirmed by the report [46], which includes classifying common features of both domains (Figure 9a). These include aesthetics, consistency in their use worldwide, and increased efficiency of photovoltaic cells with the same average visible light transmittance. Figure 9b shows the high photoconversion efficiency of cells based on crystalline silicon and perovskite, which was 26.1% and 25.2%, respectively. The efficiency mentioned above exceeds 25%, and similar values are also observed in the research on transparent photovoltaics. In [47], we can also find that for a perovskite solar cell, the high conversion efficiency was obtained equal to 22.6%; however, in [48], its record photoconversion efficiency was 25.7%. Moreover, the highest lab efficiency in thin film technology is 23.4% for CIGS and 21.0% for CdTe solar cells.

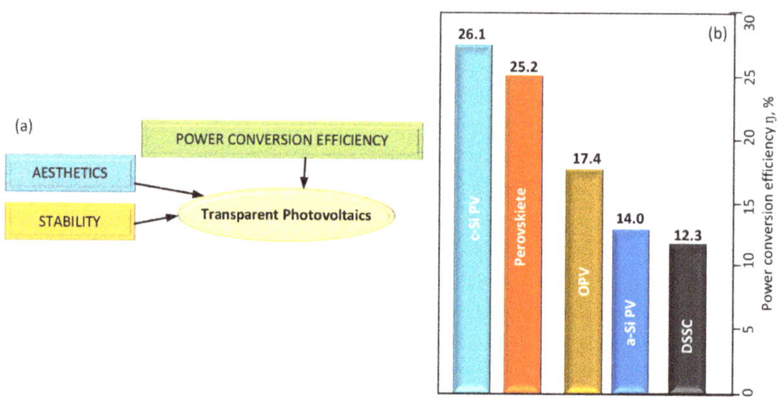

Figure 9. The perspective of the development of transparent photovoltaics: (**a**) classification of features, (**b**) classification of photoconversion efficiency of cells [46].

In [47], we find information about record lab cell efficiency for mono-crystalline and multi-crystalline silicon wafer-based technology (26.7% and 24.4%). In 2040, it is predicted that passivated emitter and rear Cells (PERC) made from c-Si and PV cells with aluminium back surface field (BSF) will be among the most prevalent solar cells in mass production using thick-film technology, which is used to produce electronic devices/modules [47].

In the case of optical tests, transmittance should be taken into account when analyzing the efficiency of the photovoltaic conversion of a cell. The transmittance should be considered an essential parameter with the photovoltaic conversion efficiency since the efficiency changes with it. Electronically conductive transparent layers are generally obtained based on compounds of indium, tin and zinc doped with tin [48,49]. Indium Tin Oxide (ITO) and Aluminium-doped Zinc Oxide (AZO) will be used in this work; therefore, they will be discussed later. The high transparency of 85–90%, conductivity above 1000 S/cm, charge carrier density of $1{\cdot}10^{21}$ carriers/cm^3, and mobility of 10–100 cm^2/Vs are the essential beneficial properties of ITO commonly known in the literature [50–52]. The ITO is often used in organic electroluminescence devices (for instance, thin-film transparent electrodes of liquid crystal displays (LCDs) [49,53–57]. The commercially available ITO targets with different doping contents can be used for the ITO deposition at relatively low temperatures [54,58–60]. Al-doped zinc oxide is one of the various indium-free materials, which has been known as one of the most promising materials for replacing ITO in electronic devices, because of its abundance, non-toxicity, high conductivity, high transparency in the visible region, and low costs [61,62]. Zinc oxide (ZnO) thin film, due to its wide range of electrical and optical properties, has become a technologically important material. It has been doped with n-type dopants such as Al, In, Ga, Mg, and B to improve ZnO electrical conductivity, optical properties, and thermal stability by providing the extra electron [63]. AZO thin films have been fabricated by various techniques, such as atomic layer deposition and pulsed laser deposition [64], magnetron sputtering, chemical vapor deposition [65], and solution processing [66–69]. Fluorine-doped tin oxide (FTO) thin film has been used mainly for electronics devices making technologies, for instance, window layers in solar cells [70], gas sensor devices [71], substrates for electro-deposition [72], and transparent contact in optoelectronic and so on. For photovoltaic applications, TCO can be applied as front electrodes of thin film solar cells because of their excellent electrical properties (4.3×10^{-4} ω.cm 4.3×10^{-4} ω.cm) combined with high transmission properties (86%) [73]. The paper [74] presented a concept of preparing TCO nanocomposite by combining TCOs and nanoparticles. This example provides guidelines for designing hazy electrodes as light management structures in thin film photovoltaics. FTO is characterized in various technology fields because of its high electrical conductivity, chemical and thermal stability, and optical transparency in the visible range (74). For example, in paper [75], a sol-gel-spin-coating method was used to create a reproducible, scalable, stable, transparent, and conducting tin fluorine-doped indium oxide thin film (FTIO thin film). The method reduced sheet resistance by more than 53% ($\leq 7 \pm 1$ Ω/sq) of the initial sheet resistance. Pulsed laser deposition [76], chemical vapor deposition (CVD) [77], and sol-gel and spray pyrolysis deposition (SPD) [78]—are just some of the methods used to prepare TCO [79]. Figure 10 shows an example of a PV structure and the use of TCO in this structure.

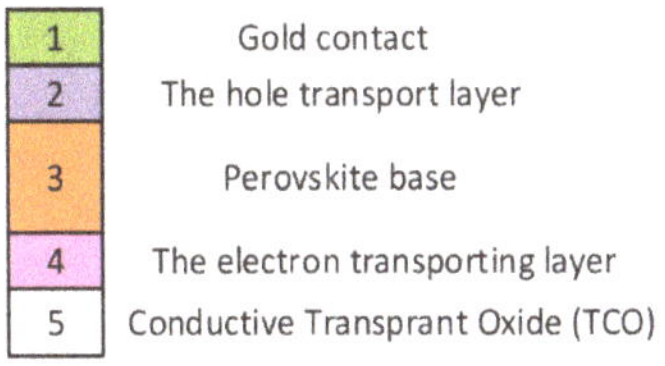

Figure 10. An example of a planar structure of a perovskite-based photovoltaic cell [1].

The issue of TCO was analyzed based on data from the Elsevier (Science Direct), Web of Science, and Scopus databases [80–82]. Based on Table 3, it can be concluded that the

Elsevier database recorded the highest number of published papers. The Scopus database was second, and WOS was third regarding issues analyzed (state of knowledge as of 7.09.22).

Table 3. List of issued publications on general topics according to three databases [80–82].

Issue Searched	Year		WOS	SCOPUS	ELSEVIER
Transparent Conductive Oxides	2012–2022	2382	814	4172	814

Table 4 summarizes the results of specific issues in the two databases, WOS and Scopus, as of 2020. Analyzing the results obtained, it can be concluded that a higher number of published papers related to issues related to SnO_2 is contained in the WOS database than in Scopus, while a comparable number of publications was found for the issue of "*AZO and* In_2O_3" for both databases is contained in the WOS database and Scopus. The issue "FTO and In_2O_3" contains the WOS database than Scopus (state of knowledge as of 5 April 2022).

Table 4. List of publications issued according to two databases with details [80,81].

WOS	SCOPUS	WOS	SCOPUS	WOS	SCOPUS	WOS	SCOPUS
AZO and In_2O_3		FTO and In_2O_3		In_2O_3		TCO and In_2O_3	
6	6	38	20	9.116	5.710	561	376
AZO and SnO_2		FTO and SnO_2		SnO_2		TCO and SnO_2	
22	36	1478	401	34.258	25.305	579	516

Some discrepancies in the scientific circulation of published research results in scientific journals in the aforementioned bibliographic databases may be due to the emergence of an increasing number of scientific journals that do not consider the low quality of non-scientific publications. Thanks to the development of information technology, bibliographic lists have been transferred to the virtual space, which has given them a new meaning [83].

5. Summary

External operating parameters characterizing electronic instruments (for example, open-circuit voltage (V_{oc}), short-circuit current (I_{sc}), filling factor (FF), and photovoltaic conversion efficiency (E_{ff}) are the values depending on the material and physical parameters of their separate layers and structural elements [84,85]. The elements and functional layers of an electronic device based on crystalline silicon (c-Si) include an impurity layer forming the emitter of the device, a passivating layer, an anti-reflective coating (ARC), and a front and rear electrode [78,85]. The electronic device's final operating parameters also determine the base material's parameters: the starting crystalline silicon wafer, the essential properties of resistivity and the lifetime of minority charge carriers. Consequently, only the mutual combination of the parameters of the base material, the parameters of the layers, and the functional elements of the instrument allows for obtaining an electronic device with the highest quantum efficiency value available for a given set of output parameters [78,85].

Production technologies and the application of surface layers have a stable position and are regarded as basic knowledge in the field of material engineering. In light of the analysis of the literature review, the authors set out two new directions worthy of attention for further research. Further work will require, as appropriate: determining the properties of thin TCO layers applied on silicon substrates by the ALD method for application as conductive emitter coatings in engineering structures, development of engineering structures based on silicon substrates with TCO layers and front contacts, development of a technology for applying a thin layer of TCO on a silicon substrate. According to the requirement of EU Directive 2009/28/EC of the European Parliament and the Council, the increase in the use of renewable energy sources (RES) in the final energy balance will be up to 20% in 2030. Thus, the direction of actions taken in the implementation doctorate is consistent with the implementation of the Polish Energy Policy until 2030 [10].

In sum, the implementation of significant achievements in the coming decade is in line with the fundamental guidelines for the development of photovoltaics as a result of [10,46,78] for the development of guidelines for metallization processes and TCO layers deposited by the ALD method.

Author Contributions: Conceptualization, M.M.-S., writing—original draft preparation M.M.-S. and writing—review and editing, M.M.-S., A.C. and R.R.; supervision, A.C. All authors have read and agreed to the published version of the manuscript.

Funding: Publication supported by the Rector's pro-quality grant. Silesian University of Technology, grant number 10/050/RGJ22/1028.

Institutional Review Board Statement: Not applicable.

Informed Consent Statement: Not applicable.

Data Availability Statement: Not applicable.

Conflicts of Interest: The authors declare no conflict of interest.

References

1. Znajdek, K.; Sibiński, M. *Postępy w Fotowoltaice*; Wydawnictwo Naukowe PWN: Warsaw, Poland, 2021; pp. 60–74. (In Polish)
2. Linie, Których Nie Widać Zrewolucjonizują Branżę Fotowoltaiczną. Available online: https://magazynfotowoltaika.pl/linie-ktorych-nie-widac-zrewolucjonizuja-branze-fotowoltaiczna (accessed on 22 August 2022).
3. VDMA. *International Technology Roadmap for Photovoltaics (ITRPV)*, 12th ed.; VDMA: Frankfurt, Germany, 2020.
4. SolarPower Europe. Available online: https://www.solarpowereurope.org (accessed on 22 August 2022).
5. Dobrzański, L.A.; Pakuła, D.; Staszuk, M. Chemical Vapor Deposition in Manufacturing. In *Handbook of Manufacturing Engineering and Technology*; Nee, A.Y.C., Ed.; Springer: Berlin/Heidelberg, Germany, 2014; Chapter 30; pp. 2755–2803.
6. Drive, P.H. Laser systems and Processes within Next Generation Photovoltaic Manufacturing Equipment. In Proceedings of the 4th Photovoltaic Manufacturing Technology Conference, Stuttgart, Semicon Europe, Stuttgart, Germany, 8–9 October 2008.
7. Musztyfaga-Staszuk, M. SLS: One of the Modern Technologies of Laser Surface Treatment. *Int. J. Thermophys.* **2017**, *38*, 130. [CrossRef]
8. Mertens, K. *Photovoltaics: Fundamentals, Technology, and Practice*, 2nd ed.; Wiley-Blackwell: Columbia, MD, USA, 2018.
9. Cui, Z. *Printed Electronics: Materials, Technologies and Applications*; Higher Education Press: Beijing, China, 2016.
10. Directive 2009/28/EC of the European Parliament and of the Council. Available online: https://eur-lex.europa.eu/LexUriServ/LexUriServ.do?uri=OJ:L:2009:140:0016:0062:en:PDF (accessed on 1 August 2022).
11. Metz, A.; Fischer, M.; Trube, J. Crystalline silicon Technology—Current Status and Outlook. In *International Technology Roadmap for Photovoltaics (ITRPV)*, 8th ed.; PV Manufacturing in Europe Conference Brussels; VDMA: Frankfurt, Germany, 2017.
12. Mat Desa, M.K.; Sapeai, S.; Azhari, A.W.; Sopian, K.; Sulaiman, M.Y.; Amin, N.; Zaidi, S.H. Silicon back contact solar cell configuration: A pathway towards higher efficiency. *Renew. Sustain. Energy Rev.* **2016**, *60*, 1516–1532. [CrossRef]
13. Kopecek, R.; Koduvelikulathu, L.J.; Cabrera, E.; Rudolph, D.; Buck, T. The technology with highest efficiency gain potential for c-Si cells. *Photovolt. Int.* **2015**, *4*, 52–60.
14. Ralph, E.L. Recent advancements in low-cost solar cell processing. In *Conference Record, Proceedings of the 11th IEEE Photovoltaic Specialists Conference, Scottsdale, AZ, USA, 6–8 May 1975*; Institute of Electrical and Electronic Engineers: Piscataway, NJ, USA, 1975; pp. 315–316.
15. Burgelman, M. Thin film solar cells by screen printing technology. In Proceedings of the Workshop Microtechnology and Thermal Problems in Electronics, Lodz, Poland, 21–27 September 1998; pp. 129–135.
16. Shanmugama, V.; Cunnusamy, J.; Khanna, A.; Boreland, B.M.; Mueller, T. Optimisation of Screen-Printed Metallisation for Industrial High-Efficiency Silicon Wafer Solar Cells. *Energy Procedia* **2013**, *33*, 64–69. [CrossRef]
17. Hamer, D.W.; Biggers, J.V. *Technologia Układów Scalonych Grubowarstwowych*; WNT: Warsaw, Poland, 1976. (In Polish)
18. Muszyfaga-Staszuk, M. *New Copper-Based Composites for the Production of Silicon Photovoltaic Cells*; WNT: Warsaw, Poland, 2019; ISBN 978-83-7880-626-4. (In Polish)
19. Hannebauer, H.; Schimanke, S.; Falcon, T.; Altermatt, P.P.; Dullweber, T. Optimised stencil print for low Ag paste consumption and high conversion efficiencies. *Energy Procedia* **2015**, *67*, 108–115. [CrossRef]
20. Caballero, L.J. Contact Definition in industrial silicon solar cells. In *Solar Energy*; Rugescu, R.D., Ed.; IntechOpen: London, UK, 2010; pp. 375–398.
21. Kofron, V.K. Photovoltaic Cell with Junction-Free Essentially-Linear Connections to Its Contacts. U.S. Patent 4,153,907, 8 May 1979.

22. Neuhaus, A.R.; Bultman, J.H.; Tip, A.C.; Sinke, W.C. Metallisation patterns for interconnection through holes. *Sol. Energy Sol. Cells* **2001**, *65*, 347–353.
23. Alemán, M.; Streek, A.; Regenfuβ, P.; Mette, A.; Ebert, R.; Exner, H.; Glunz, S.W.; Willeke, G. In Laser micro-sintering as a new metallisation technique for silicon solar cells. In Proceedings of the 21st European Photovoltaic Solar Energy Conference, Dresden, Germany, 4–8 September 2006.
24. Edwards, M. Efficiencies of 22% at low cost: The future of mass-produced laser-doped selective emitter solar cells. *Photovolt. Int.* **2011**, *1*, 56–62.
25. Hendell, R. Laser Applications in Solar Cell Manufacturing. In *Focus: Laser Micro-Processing*; Wiley-VCH: Weinheim, Germany, 2008.
26. Dross, F.; Van, K.E.; Allebe, C.; Van der Heide, A.; Szlufcik, J.; Agostinelli, G.; Choulat, P.; Dekkers, H.F.W.; Beaucarne, G. Impact of Rear-Surface Passivation on MWT Performances. In *Conference Record, Proceedings of the IEEE 4th World Conference on Photovoltaic Energy Conversion, Waikoloa, HI, USA, 7–12 May 2006*; Institute of Electrical and Electronic Engineers: Piscataway, NJ, USA, 2006.
27. Kerscharer, E.V.; Beaucarne, G. Back-contact Solar Cells: A review. Progress in Photovoltaic: Research and application. *Prog. Photovolt. Appl.* **2005**, *14*, 107–123. [CrossRef]
28. Gautero, L.; Hofmann, M.; Rentsch, J.; Lemke, A.; Mack, S.; Seiffe, J.; Nekarda, J.; Biro, D.; Wolf, A.; Bitnar, B.; et al. All-screen-printed 120-µm-thin large-area silicon solar cells applying dielectric rear passivation and laser-fired contacts reaching 18% efficiency. In Proceedings of the 34th IEEE Photovoltaic Specialists Conference (PVSC), Philadelphia, PA, USA, 7–12 June 2009.
29. Neuhaus, D.H.; Münzer, A. Industrial silicon wafer solar cells. *Adv. Opto Electron.* **2007**, *2007*, 24521. [CrossRef]
30. Nande, A.; Raut, S.; Dhoble, S.J. Energy Materials. In *Perovskite Solar Cells*; Dhoble, S.J., Thejo Kalyani, N., Vengadaesvaran, B., Arof, A.K., Eds.; Elsevier: Amsterdam, The Netherlands, 2021; Chapter 9; pp. 249–281.
31. Yang, D.; Zhang, X.; Hou, Y.; Wang, K.; Ye, T.; Yoon, J.; Wu, C.; Sanghadasa, M.; Liu, S.; Priya, S. 28.3%-efficiency perovskite/silicon tandem solar cell by optimal transparent electrode for high efficient semitransparent top cel. *Nano Energy* **2021**, *84*, 105934. [CrossRef]
32. Metallization Paste Manufacturers. Available online: https://www.enfsolar.com/directory/material/metallization_paste?tech=408 (accessed on 15 June 2022).
33. TaiyangNews Provides Update on Latest Technical and Product Solutions to Contact All Types of Solar Cells. Available online: https://taiyangnews.info/reports/market-survey-metallization-pastes-201920/ (accessed on 10 April 2022).
34. Johnson, R.W.; Hultqvist, A.; Bent, S.F. A brief review of atomic layer deposition: From fundamentals to applications. *Mater. Today* **2014**, *17*, 236–246. [CrossRef]
35. Staszuk, M.; Pakuła, D.; Reimann, Ł.; Król, M.; Basiaga, M.; Mysłek, D.; Kři, A. Structure and Properties of ZnO Coatings Obtained by Atomic Layer Deposition (ALD) Method on a Cr-Ni-Mo Steel Substrate Type. *Materials* **2020**, *13*, 4223. [CrossRef] [PubMed]
36. Lujala, V.; Skarp, J.; Tammenmaa, M.; Suntola, T. Atomic layer epitaxy growth of doped zinc oxide thin films from organometals. *Appl. Surf. Sci.* **1994**, *82/83*, 34–40. [CrossRef]
37. Osadzanie Warstw Atomowych (ALD). Available online: http://www.eagle-regpot.eu/EfI/index.php?pid=33 (accessed on 1 August 2022).
38. Wojcik, A.; Kiecana, M.; Kopalko, K.; Godlewski, M.; Guziewicza, E.; Yatsunenko, S.; ÃLusakowska, E.; Minikayev, R.; Paszkowicza, W.; Swiatek, K.; et al. Magnetic, structural, and optical properties of low-temperature ZnMnO grown by atomic layer epitaxy; Proceedings of the XXXIV International School of Semiconducting Compounds. *Acta Phys. Pol. A* **2005**, *108*, 915–921. [CrossRef]
39. Robertson, J. High dielectric constant oxides. *Eur. Phys. J. Appl. Phys.* **2004**, *28*, 265–291. [CrossRef]
40. Boryło, P.; Szindler, M.; Lukaszkowicz, K. *Various Applications of Multifunctional Thin Films with Specific Properties Deposited by the ALD Method*; Trans Tech Publications: Wollerau, Switzerland, 2019; pp. 111–123.
41. Provine, J.; Rincon, M. *Atomic Layer Deposition: Introduction to the Theory and Cambridge Nanotech Savannah & Fiji*; Stanford University: Stanford, CA, USA, 2012.
42. Poortmans, J.; Pieters, P.; Baert, K. Exploiting the microelectronics toolbox to boost Si PV manufacturing. *Photovolt. Int.* **2011**, *3*, 102–110.
43. Grochowski, J.; Guziewicz, M.; Borysiewicz, M. Analiza transmisji optycznej półprzewodnikowych warstw NiO osadzanych metodą magnetronowego rozpylania katodowego. *Elektron. Konstr. Technol. Zastos.* **2011**, *52*, 7. (In Polish)
44. Lee, K.; Kim, N.; Kim, K.; Um, H.-D.; Jin, W.; Choi, D.; Park, J.; Park, K.J.; Lee, S.; Seo, K. Neutral-Colored Transparent Crystalline Silicon Photovoltaics. *Joule* **2020**, *4*, 235–246. [CrossRef]
45. Fraunhofer Institute for Solar Energy Systems (ISE). *Photovoltaics Report*; Prepared with support of PSE Conferences &Consulting GmbH; Fraunhofer Institute for Solar Energy Systems: Freiburg im Breisgau, Germany, 2021.
46. NREL-Led Research into Perovskite-Silicon Tandem Cells Shows New Path to Take. Available online: https://www.nrel.gov/news/program/2020/nrel-led-research-into-perovskite-silicon-tandem-cells-shows-new-path.html (accessed on 11 August 2022).
47. Kryłow, J.; Oleński, J.; Sawicki, Z.; Tumański, A. *Domieszkowanie Półprzewodników, Procesy Technologiczne w Elektronice Półprzewodnikowej*; WN: Warsaw, Poland, 1980. (In Polish)
48. Vlooswijk, A. Growth & Characterisation of p-Type Transparent Oxide Semiconductors. Master's Thesis, Applied Physics, Inorganic Materials Science, Faculty of Science and Technology, University of Twente, Enschede, The Netherlands, February 2005.
49. Klein, E.; Hubert, K.; Paul, O.; Ruther, P. Low-temperature plasma annealing of sputtered indium tin oxide for transparent and conductive thin-films on glass and polymer substrates. *Thin Solid Films* **2020**, *693*, 137715. [CrossRef]

50. Hamberg, I.; Granqvist, C.G. Evaporated Sn-doped In_2O_3 films: Basic optical properties and applications to energy-efficient windows. *J. Appl. Phys.* **1986**, *60*, R123–R159. [CrossRef]
51. Laist, J.W. *Copper, Silver and Gold*, 2nd ed.; D. van Nostrand Company: Toronto, ON, Canada, 1954.
52. Ross, R.B. Copper Cu. In *Metallic Materials Specification Handbook*; Springer: Boston, MA, USA, 1992.
53. Bartsch, J.; Mondon, A.; Bayer, K.; Schetter, C.; Hörteis, M.; Glunz, S.W. Quick Determination of Copper-Metallization Long-Term Impact on Silicon Solar Cells. *J. Electrochem. Soc.* **2010**, *157*, H942. [CrossRef]
54. Minami, T. Present status of transparent conducting oxide thin-film development for Indium-Tin-Oxide (ITO) substitutes. *Thin Solid Films* **2008**, *516*, 5822–5828. [CrossRef]
55. Tak, Y.H.; Kim, K.B.; Park, H.G.; Lee, K.H.; Lee, J.R. Criteria for ITO (indium-tin-oxide) thin film as the bottom electrode of an organic light-emitting diode. *Thin Solid Films* **2002**, *411*, 12–16. [CrossRef]
56. Song, P.K.; Akao, H.; Kamei, M.; Shigesato, Y.; Yasui, I. Preparation and crystallisation of tin-doped and undoped amorphous indium oxide films deposited by sputtering. *Jpn. J. Appl. Phys.* **1999**, *38*, 5224–5226. [CrossRef]
57. Nam, E.; Kang, Y.; Son, D.; Jung, D.; Hong, S.; Kim, Y.S. Electrical and surface properties of indium tin oxide (ITO) films by pulsed DC magnetron sputtering for organic light-emitting diode as anode material. *Surf. Coat. Technol.* **2010**, *205*, S129–S132. [CrossRef]
58. Minami, T.; Sonohara, H.; Kakumu, T.; Takata, S. Physics of very thin ITO conducting films with high transparency prepared by DC magnetron sputtering. *Thin Solid Films* **1995**, *270*, 37–42. [CrossRef]
59. Ellmer, K. Past achievements and future challenges in the development of optically transparent electrodes. *Nat. Photonics* **2012**, *6*, 809. [CrossRef]
60. Sun, L.; Grant, J.T.; Jones, J.G.; Murphy, N.R. Tailoring electrical and optical properties of Al-doped ZnO thin films grown at room temperature by reactive magnetron co-sputtering: From bandgap to near-infrared. *Opt. Mater.* **2018**, *84*, 146–157. [CrossRef]
61. Ellmer, K.; Bikowski, A. Intrinsic and extrinsic doping of ZnO and ZnO alloys. *J. Phys. D Appl. Phys.* **2016**, *49*, 413002. [CrossRef]
62. Wu, Y.; Potts, S.; Hermkens, P.; Knoops, H.; Roozeboom, F.; Kessels, W. Enhanced doping efficiency of Al-doped ZnO by atomic layer deposition using dimethyl aluminium isopropoxide as an alternative aluminium precursor. *Chem. Mater.* **2013**, *25*, 4619–4622. [CrossRef]
63. Knoops, H.C.; Loo, B.W.; Smit, S.; Ponomarev, M.V.; Weber, J.-W.; Sharma, K.; Kessels, W.M.; Creatore, M. Optical modeling of plasma-deposited ZnO films: Electron scattering at different length scales. *J. Vac. Sci. Technol.* **2015**, *33*, 021509. [CrossRef]
64. Zhang, H.; Li, X.; Fang, Z.; Yao, R.; Zhang, X.; Deng, Y.; Lu, X.; Tao, H.; Ning, H.; Peng, J. Highly conductive and transparent AZO films fabricated by PLD as source/drain electrodes for TFTs. *Materials* **2018**, *11*, 2480. [CrossRef] [PubMed]
65. Shahid, M.; Deen, K.; Ahmad, A.; Akram, M.; Aslam, M.; Akhtar, W. Formation of Al-doped ZnO thin films on glass by sol-gel process and Characterisation. *Appl. Nanosci.* **2016**, *6*, 235–241. [CrossRef]
66. Islam, M.R.; Rahman, M.; Farhad, S.; Podder, J. Structural, optical and photocatalysis properties of sol-gel deposited Al-doped ZnO thin films. *Surf. Interfaces* **2019**, *16*, 120–126. [CrossRef]
67. Nasiri, M.; Rozati, S. Muscovite mica as a flexible substrate for transparent conductive AZO thin films deposited by spray pyrolysis. *Mater. Sci. Semicond. Process.* **2018**, *81*, 38–43. [CrossRef]
68. Stapiński, T.; Godlewski, M.; Jakubowska, M.; Marszałek, K.; Pietruszka, R.; Panek, P.; Soliński, B.; Soliński, I.; Turoń, K.; Wróblewski, G. *Materiały i Metody Optymalizacji Budowy Ogniw i Paneli Fotowoltaicznych*; AR TOP: Kraków, Poland, 2014; pp. 15–77.
69. Luque, A.; Hegedus, S. (Eds.) *Handbook of Photovoltaic Science and Engineering*; John Wiley & Sons, Ltd.: New York, NY, USA, 2003.
70. Chae, J.; Kim, D.Y.; Kim, S.; Kang, M. Photovoltaic efficiency on dye-sensitised solar cells (DSSC) assembled using Ga-incorporated TiO_2 materials. *J. Ind. Eng. Chem.* **2010**, *16*, 906–911. [CrossRef]
71. Hossein-Babaei, F.; Amini, A. A breakthrough in gas diagnosis with a temperature-modulated generic metal oxide gas sensor. *Sens. Actuators B Chem.* **2012**, *166–167*, 419–425. [CrossRef]
72. Khelladi, M.; Mentar, L.; Boubatra, M.; Azizi, A.; Kahoul, A. Early stages of cobalt electro-deposition on FTO and n-type Si substrates in sulphate medium. *Mater. Chem. Phys.* **2010**, *122*, 449–453. [CrossRef]
73. Muthukumar, A. Fluorine-doped tin oxide (FTO) thin film as transparent conductive oxide (TCO) for photovoltaic applications. *AIP Conf. Proc.* **2013**, *1512*, 710.
74. Nazeeruddin, M.K.; Baranoff, E.; Grätzel, M. Dye-sensitised solar cells: A brief overview. *Sol. Energy* **2011**, *85*, 1172–1178. [CrossRef]
75. Kadam, A.N.; Chowdhury, S.R.; Bathula, C.; Kumar, N.; Kumar, V.; Jha, M.K.; Lee, S.-W.; Misra, M. A novel reduction approach for fabrication of transparent conducting fluorine and tin-doped indium oxide thin film with low sheet resistance. *Ceram. Int.* **2022**, *48*, 29307–29313. [CrossRef]
76. Chen, Z.; Lai, J.; Shek, C. Multifractal spectra of scanning electron microscope images of SnO_2 thin films prepared by pulsed laser deposition. *Phys. Lett.* **2005**, *345*, 218–223. [CrossRef]
77. Fang, T.-H.; Chang, W.-J. Effect of freon flow rate on tin oxide thin films deposited by chemical vapour deposition. *Appl. Surf. Sci.* **2003**, *220*, 175–180. [CrossRef]
78. Moholkar, A.; Pawar, S.; Rajpure, K.; Bhosale, C. Effect of solvent ratio on the properties of highly oriented sprayed fluorine-doped tin oxide thin films. *Mater. Lett.* **2007**, *61*, 3030–3036. [CrossRef]
79. Napi, M.L.M.; Maarof, M.F.; Soon, C.F.; Nayan, N.; Fazli, F.I.M.; Hamed, N.K.A.; Mokhtar, S.M.; Seng, N.K.; Ahmad, M.K.; Suriani, A.B.; et al. Fabrication of fluorine-doped tin oxide (FTO) thin films using spray pyrolysis deposition method for transparent conducting oxide. *J. Eng. Appl. Sci.* **2016**, *11*, 8800–8840.

80. Web of Science (Clarivate). Available online: https://www.webofscience.com/ (accessed on 7 September 2022).
81. SCOPUS. Available online: https://www.scopus.com (accessed on 7 September 2022).
82. Elsevier (Science Direct). Available online: https://www.sciencedirect.com/ (accessed on 7 September 2022).
83. Indeksowanie Czasopism w Referencyjnych Bazach Danych. Available online: https://depot.ceon.pl/bitstream/handle/123456789/15614/Aneta_Drabek_Indeksowanie_czasopism_w_referencyjnych_bazach_danych.pdf?sequence=1&isAllowed=y (accessed on 6 September 2022).
84. Boxwell, M. *Solar Electricity Handbook: A Simple, Practical Guide to Solar Energy—Designing and Installing Solar Photovoltaic Systems*, 2022 ed.; Greensteram Publishing: Birmingham, UK, 2022.
85. Zalewski, E.F.; Geist, J. Solar cell spectral response characterisation. *Appl. Opt.* **1979**, *18*, 3942–3947. [CrossRef]

energies

MDPI

Article

Studies on Water–Aluminum Scrap Reaction Kinetics in Two Steps and the Efficiency of Green Hydrogen Production

Ansis Mezulis [1], Christiaan Richter [2], Peteris Lesnicenoks [1], Ainars Knoks [1], Sarunas Varnagiris [3], Marius Urbonavicius [3], Darius Milcius [3] and Janis Kleperis [1,*]

[1] Institute of Solid State Physics, University of Latvia, LV-1063 Riga, Latvia; ansis.mezulis@cfi.lu.lv (A.M.); peterisl@cfi.lu.lv (P.L.); ainars.knoks@cfi.lu.lv (A.K.)

[2] Faculty of Industrial Engineering, Mechanical Engineering and Computer Science, University of Iceland, 102 Reykjavik, Iceland; cpr@hi.is

[3] Center for Hydrogen Energy Technologies, Lithuanian Energy Institute, 3 Breslaujos, 44403 Kaunas, Lithuania; sarunas.varnagiris@lei.lt (S.V.); marius.urbonavicius@lei.lt (M.U.); darius.milcius@lei.lt (D.M.)

* Correspondence: janis.kleperis@cfi.lu.lv

Abstract: This work aims to explain aluminum hydrolysis reaction kinetics based on a properly chosen theoretical model with machined aluminum waste chips as well as alkali solutions up to 1M as a promoter and to estimate the overall reaction profit. The purpose of this work is to assess the optimal alkali concentration in the production of small- and medium-scale green hydrogen. To obtain results with better accuracy, we worked with flat Al waste chips, because a flat surface is preferable to maximally increase the time for the created hydrogen bubbles to reach the critical gas pressure. Describing the reaction kinetics, a flat shape allows for the use of a planar one-dimensional shrinking core model instead of a much more complicated polydisperse spheric shrinking core model. We analyzed the surface chemical reaction and mass transfer rate steps to obtain the first-order rate constant for the surface reaction and the diffusion coefficient of the aqueous reactant in the byproduct layer, respectively. We noted that measurements of the diffusion coefficient in the byproduct layer performed and discussed in this paper are rare to find in publications at alkali concentrations below 1M. With our reactor, we achieved a H_2 yield of 1145 mL per 1 g of Al with 1M NaOH, which is 92% of the theoretical maximum. In the estimation of profit, the authors' novelty is in paying great attention to the loss in alkali and finding a crucial dependence on its price. Nevertheless, in terms of consumed and originated materials for sale, the conversion of aluminum waste material into green hydrogen with properly chosen reaction parameters has positive profit even when consuming an alkali of a chemical grade.

Keywords: aluminum waste; aluminum hydrolysis; alkali promoter; reaction rate; hydrogen production efficiency

1. Introduction

With the world's primary aluminum production of 65 million metric tons in 2020, the worldwide issue of residual aluminum waste from industry, contrary to conventional assumptions, is not resolved. Due to the content of domestic waste and the amount of aluminous residual byproducts produced by the primary and secondary Al industries, aluminum can be found in landfills in substantial quantities. In general, the byproduct of the aluminum recycling industry and the Bayer process is known as "aluminum dross". In most regions Al dross is not classified as a hazardous waste material, and it can be stored in landfills without any pretreatment. Nevertheless, Al-rich waste in landfills is a serious hazard. Aluminum, stored in landfills, may come into contact with a kind of water source after some time, including landfill leachate with a pH < 7. Such liquids, especially those with a pH < 7, can react with landfill aluminum. As a result, pockets of concentrated hydrogen appear, and their growing gaseous pressure in landfills poses a risk of combustion. Subsurface landfill fires can be rather dangerous and destructive to landfill systems because they are hard to detect, and visible smoke may not arise. The monitoring and control of potentially hazardous wastes demand a lot of money that can be assigned to research on and the development of effective methods with which to deal with aluminous waste [1,2].

The findings and reported results of various studies on dealing with aluminum waste are always dependent on the region in which R&D work is carried out. Some of the pioneering authors on this subject, such as Martinez et al. from Mexico, in 2005, reported a hydrogen gas yield of 0.049 moles from each gram of aluminum waste cans with a 2M NaOH promoter [3], with this being a decent 88% of the maximum yield of pure aluminum. Hiraki and Akiyama from Japan, in 2009, developed a system by which gaining energy from aluminum waste containing down to 15% Al is still cost-effective [4]. Instead of using chemical promoters, such as hydroxides, oxides, or salts, the authors of [5] dealt with ball milling as a pretreatment and carried out the reaction in hot water. In [6], the difference between Al dross obtained from an aluminum recycling facility (RD) and Al dross directed to landfills (LD) was studied. The authors reported that, with NaOH as the promoter, the RD and LD samples generated 0.50 and 0.15 L of H_2 per 1g of Al, respectively. The maximum flow rates were measured as 2 L/min and 0.8 L/min, respectively. The authors claimed that, on a large scale, this process can be cost-effective; moreover, the formation of bayerite and gibbsite has additional market value.

In terms of the hydrogen production concept, obtaining H_2 from the reaction of certain metals with water has been a subject of investigation for many years. Among the number of metals that, under certain circumstances, can react with aqueous solutions to produce hydrogen, aluminum and its alloys are put forward as the most suitable materials for the development of hydrogen production in the future [7]. Aluminum can be transported and stored in a much simpler, safer, and cheaper way in comparison to the end product hydrogen. Compared with the chemical rival sodium borohydride, aluminum is much less expensive and more stable under normal conditions. The price of aluminum powder is more than 10 times lower than that of sodium borohydride [7,8]. Some authors, e.g., Wang et al., have proven that H_2 produced by an aluminum chemical reaction has a purity valid for a PEM fuel cell. A mini-type hydrogen generator from 25 wt.% Al alloy strips provides a H_2 generation rate of about 38 mL/min at a 0.01 mL/s dropping rate of a sodium hydroxide solution [9]. Even though empirical evidence on hydrogen production is substantial with controlled Al alloys and some types of processed Al waste, the reaction model may vary and is not completely clear. This work aims to explain the reaction mechanism based on a properly chosen theoretical model and experimental findings with flat Al chips and alkali solutions up to 1M, and to estimate the overall reaction profit. The purpose and main goal of this work is to find out from obtained data the optimal alkali concentration in the production of small- and medium-scale green hydrogen by taking into account the loss in alkali.

2. Materials and Methods

2.1. *Aluminum Surface and Its Reaction with Water*

Aluminum is known as a highly reactive chemical element that reacts with ambient oxygen to create a protective coating. A piece of aluminum, placed into a solution, already has a coating of aluminum oxide or alumina: Al_2O_3. Aluminum oxide reacts with water at a moderate temperature to produce a boehmite AlOOH layer. This is the so-called induction step [10]:

$$Al_2O_3 + H_2O \rightarrow 2AlOOH, \quad (1)$$

At the induction step the boehmite layer grows, and the diffusion of OH^- ions through the AlOOH layer is assumed. As a result, gaseous hydrogen bubbles appear at the $Al{:}Al_2O_3$ interface:

$$6AlOOH + 2Al \rightarrow 4Al_2O_3 + 3H_2\uparrow, \quad (2)$$

The thickness of the surface passive oxide layer, Al_2O_3, on each aluminum particle is several nanometers, weakly dependent on the size of a spherical Al particle [11]. In contrast, the time required for the hydration of the oxide film has no significant difference for various Al powders. The difference in the induction time mainly comes from the accumulation of H_2 molecules in gaseous bubbles to reach the critical gas pressure.

From a physical point of view, each gaseous H_2 bubble has a critical pressure that the hydrated oxide film can withstand, as shown in Figure 1a. When, by proceeding with the chemical reaction, the gas pressure in a H_2 bubble exceeds its critical threshold, the surface film on the Al particle locally breaks, and the reaction of aluminum with water refreshes. From the vector diagram in Figure 1b, the critical gas pressure in a H_2 bubble, $P^*_{H_2}$, can be written as follows [12,13]:

$$P^*_{H_2} = P_{atm} + 2\frac{h\sigma_h cos\theta}{a} \quad (3)$$

where P_{atm} is the environmental gas pressure, h is the thickness of the hydrated oxide film, a is the radius of a H_2 bubble, and σ_h is the tensile strength of the film. Theta, θ, is the critical angle between the stress vector, $\vec{\sigma_h}$, and the normal axis to the film. Evidently, the time taken for the accumulation of H_2 quantity in the bubbles depends on their critical gas pressure. The critical gas pressure in the bubbles at the Al:Al_2O_3 interface must be lower for smaller Al particles as a smaller radius of a particle tends to reduce the total tolerable extension of a bubble on the surface. In other words, when the hydrogen gas pressure increases, a H_2 bubble expands, and for small-sized Al particles the critical angle, θ, in Figure 1b has no further deviation from 90°. Noting that, at normal conditions, the first term, P_{atm}, is more than a hundred times smaller than the second in Equation (3), a small change in the critical angle, θ, leads to a large change in the critical gas pressure, $P^*_{H_2}$.

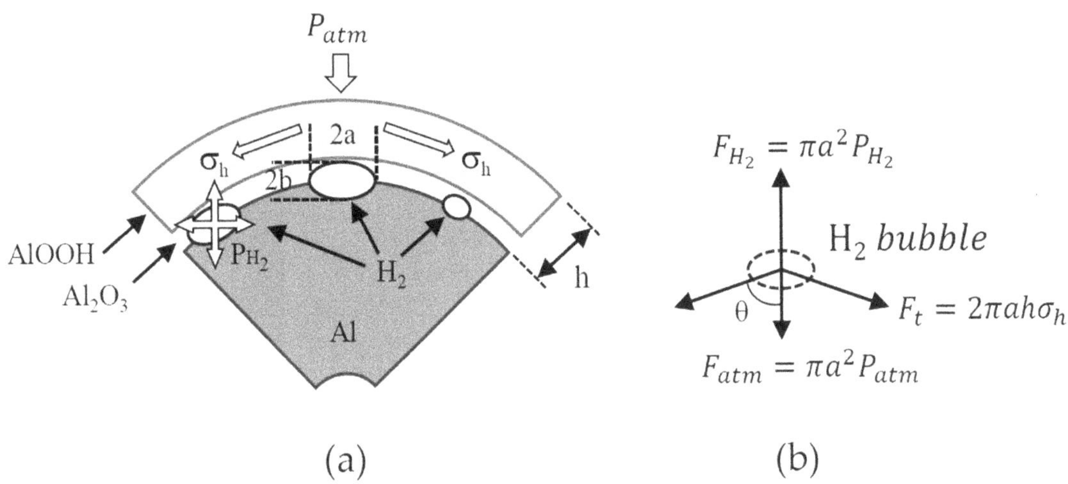

Figure 1. (**a**) A hydrogen bubble on the boundary layer of metal-hydroxide, (**b**) the acting forces on a H_2 bubble in the boundary layer [10].

Regarding the opposite consideration that a larger Al particle has a higher critical gas pressure and, therefore, larger-sized H_2 bubbles, a flat surface is preferable to increase the time for bubbles to reach the critical gas pressure. As increasing the induction time to perform the duration analysis is advantageous for the present work, instead of spheric waste particles we prefer to examine chips, i.e., flat pieces of machined aluminum waste. Describing the reaction kinetics allows us to use a planar one-dimensional shrinking core model instead of a much more complicated polydisperse spheric shrinking core model.

2.2. The Role of the Alkali Promoter

All practicable Al reactions with water are intended to use a method to eliminate the protective layer of aluminum oxide, which hinders the proceeding of the reaction with water. Between different eliminators, a chemical promoter, such as sodium or potassium hydroxide, works in a well-studied way. In the presence of sodium hydroxide as a promoter, aluminum oxide is dissolved [14]:

$$Al_2O_3 + 2NaOH + 3H_2O \rightarrow 2Na^+ + 2[Al(OH)_4]^-, \quad (4)$$

As a result, the exposed Al surface is able to react with water to form hydrogen:

$$2Al + 6H_2O \rightarrow 2Al(OH)_3 + 3H_2\uparrow \quad (5)$$

The created surface layer of aluminum hydroxide is dissolved by sodium hydroxide:

$$Al(OH)_3 + NaOH \rightarrow Na^+ + [Al(OH)_4]^- \quad (6)$$

Excluding dissociated ions, Equations (5) and (6) can be composed as follows:

$$2Al\ (s) + 6H_2O + 2NaOH\ (aq) \rightarrow 2NaAl(OH)_4\ (aq) + 3H_2\uparrow \quad (7)$$

To sum up, even at room temperature aluminum and its alloys are dissolved in an alkaline environment, which results in H_2 production. The regeneration of sodium hydroxide occurs via the decomposition of aqueous $NaAl(OH)_4$ and results in an aluminum hydroxide residue:

$$NaAl(OH)_4\ (aq) \rightarrow NaOH\ (aq) + Al(OH)_3\ (s) \quad (8)$$

There are several parameters that affect this reaction chain: the purity of aluminum material and its morphology, alkali concentration, temperature, and fresh alkaline feed rate.

2.3. Determining the Activation Energy

The activation energy of the reactant is the base parameter with which to estimate the occurrence of a chemical reaction, i.e., the amount of energy needed for molecules or atoms to break existing chemical bonds. The activation energy, E_a, is defined by the Arrhenius equation:

$$k = Ae^{-\frac{E_a}{RT}} \quad (9)$$

where k is the reaction rate constant, R is the universal gas constant, T is the absolute temperature, and A is the frequency factor constant, the value of which is rarely used. The reaction rate constant, k, here is assumed to be that of a "pure" chemical reaction, i.e., without any hindrances to the maximum reaction rate. It is assumed that the first-order rate constant for a surface reaction, k^*, at sufficient stirring can be used as k. E_a is usually derived from obtained reaction rate constants, k, at different temperatures. Applying the natural log to Equation (9) yields the following:

$$\ln(k) = \ln(A) - \frac{E_a}{RT} \quad (10)$$

With properly found values of k at various temperatures, T, plotting $\ln(k)$ vs. $1/T$ displays a straight line, known as an Arrhenius approximation. Its slope is equal to $-E_a/R$, whereas the y-intercept is equal to $\ln(A)$.

2.4. Reaction Kinetics in Two Steps

According to the ideal gas model, the hydrogen yield, α ($0 < \alpha < 1$), can be written as follows:

$$\alpha(t) = \frac{p_{eff} V_{eff}}{n_0 RT} \tag{11}$$

where p_{eff} is the effective pressure (with subtracted initial gas pressure), V_{eff} is the effective volume (with subtracted volumes, occupied by liquid and powder), and n_0 is the theoretically calculated number of H_2 moles by reacting all of the aluminum metal.

The aluminum–water reaction is a solid–liquid heterogeneous system. Its reaction dynamics are usually described through a shrinking core model. A well-known book [15] gives the formulae of shrinking cores for three geometries of reacting solid bodies: flat plates, cylinders, and spheres. With the examined sample we choose the flat plates model, the geometry of which is determined only by the half-thickness of the plate, r. In describing the reaction between a large number of small solid bodies and an aqueous solution, two main reaction steps are considered.

2.4.1. Surface Reaction Rate Step

In the first step the reaction rate is controlled by the surface chemical reaction, and the equation that describes the shrinking of a flat solid plate at isothermal conditions is as follows:

$$\frac{t}{\tau} = \alpha(t) \tag{12}$$

where t is the real time, and the characteristic time constant, τ, can be found from the following equation:

$$\tau = \frac{r\rho_{Al}}{bk^* c_{alk} M_{Al}} \tag{13}$$

where ρ_{Al} and M_{Al} are the density and molar mass of the solid (aluminum), respectively, c_{alk} is the molar concentration of the alkali aqueous promoter, and b is the stoichiometry coefficient in the reaction between solid and liquid, $n_{Al} = b \cdot n_{alk}$, being equal to 2/3 in the examined reactions, such as Equation (7). The first-order rate constant for surface reactions, k^* (mm/s), is the coefficient that is determined by processing experimental data. In [16] it is suggested to take k^* as a composition of the surface reaction rate constant, k_s, and the mass transfer constant in laminar boundary layers, k_c:

$$\frac{1}{k^*} = \frac{1}{k_s} + \frac{1}{k_c} \tag{14}$$

At sufficient stirring of aqueous alkali solutions, $k_c \gg k_s$ and $k^* \approx k_s$. In experiments, one can choose the intensity of stirring and therefore to exclude the mass transfer constant, k_c.

2.4.2. Mass Transfer Rate Step

In the second step the reaction rate is mainly controlled by the mass transfer in the byproduct layer, and for a flat solid plate at isothermal conditions the corresponding equations are as follows:

$$\frac{t}{\tau} = \alpha^2(t) \tag{15}$$

$$\tau = \frac{r^2\rho_{Al}}{2bDc_{alk} M_{Al}}, \tag{16}$$

where D (mm^2/s) is the diffusion coefficient of the aqueous reactant in the byproduct layer. Likewise, parameter k^* in the surface reaction rate step, D, is the parameter to be determined by the experimental data processing in the mass transfer rate step.

3. Materials and Methods

3.1. Experimental Sample

Flat Al chips are waste products from manufacturing aluminum constructions for insulating glass windows. We receive the chips, with an average thickness of 0.2 mm, from a Lithuanian company, "Stiklita, JSC". An initial analysis of the chips was performed on the surface morphology via a scanning electron microscope (SEM, Hitachi S-3400N, Tokyo, Japan) and on the bulk elemental composition as well as elemental mapping via energy-dispersive X-ray spectroscopy (Bruker Quad 5040, Hamburg, Germany); the results are visible in Table 1. Additionally, the crystal structure of the initial Al material was examined via an X-ray diffractometer (XRD, Bruker D8, Hamburg, Germany), Cu Kα radiation, and a Lynx Eye linear position detector (Hamburg, Germany) at 2 theta angles in the range of 20° to 70°. The results can be seen in Figure 2: intense aluminum peaks with cubic crystallographic orientations of (111), (200), and (220) at $2\Theta = 38.2°, 44.5°$, and 64.9°. The results of these analyses are compiled here in Figure 2: Alkali pellets, NaOH and KOH, have purities of >99%. The activation energy of our flat 6063 aluminum alloy chips, obtained from a temperature-dependent series via Arrhenius approximation, is 48.1 kJ/mol. We can compare this value to that given in [17], obtained by applying 1–2M NaOH alkali. The extrapolation of very small dependence on the alkali molar concentration gives E_a = 40.9 kJ/mol with a 0.02 mm-thick foil and 57.2 kJ/mol with 0.5 mm thick plates. Thus, the value of E_a in this work is in the middle, which is fully consistent regarding the different material structures of the samples.

Table 1. Elemental composition of Al chips as received.

Elemental Composition	Concentration, %
Aluminum	94.3
Magnesium	0.6
Carbon	5.0
Oxygen	0.1

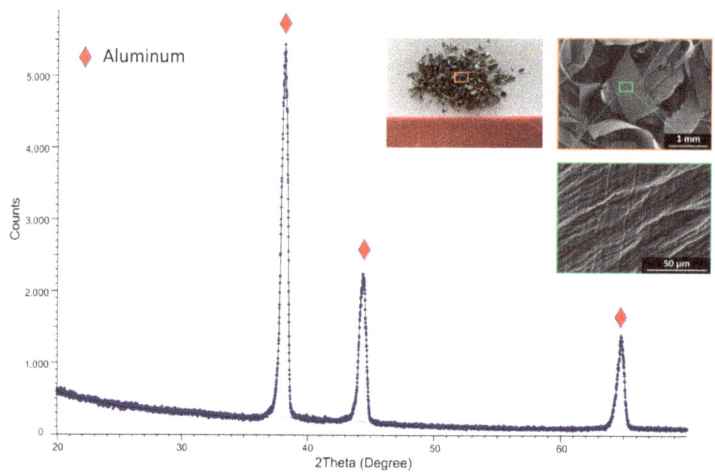

Figure 2. Initial waste Al chips: XRD graph; photo camera image; and SEM images at different magnitudes.

3.2. Experimental Setup

To investigate aluminum hydrolysis, we used a custom-made laboratory stand, shown in Figure 3. Before starting each experiment, we created a 10 Pa vacuum with a turbomolecular pump, No. 6 in Figure 3, in all volumes designed for the spreading of the produced hydrogen. A measured dose of initial waste Al chips from the container, 1, plunges into the reaction volume, 3. The magnetic stirrer is used to maintain a sufficiently low mass transfer constant in the solid–liquid laminar boundary layer. The water, required for the hydrolysis reaction, was deionized and purged by argon flow. The produced hydrogen is purified by passing through the water in 9 and spreads to the main volume, 12, where its samples can be taken for spectral analyses. The gas spectral analysis did not show deviation from expected pure hydrogen, with up to 99.98% hydrogen.

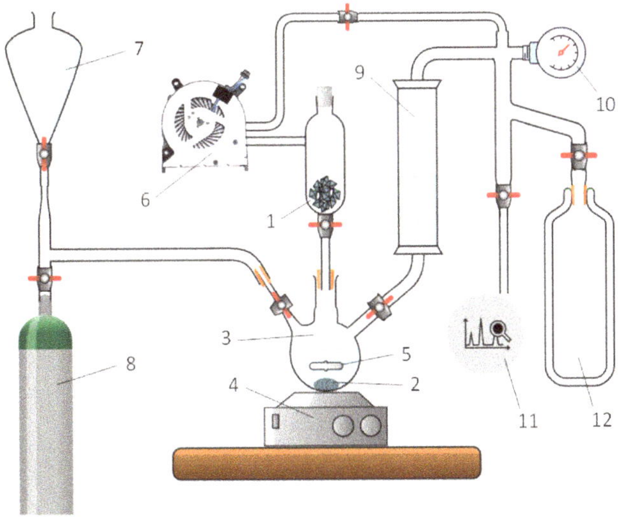

Figure 3. Experimental setup of the chemical reactor to produce hydrogen from aluminum. 1—Al material (chips); 2—$Al(OH)_3$ residue; 3—reactor volume; 4—magnetic stirrer; 5—stirrer stick; 6—turbomolecular pump; 7—water deionizer; 8—argon tank; 9—gaseous hydrogen pre-volume; 10—manometer; 11—gas spectral analyzer; and 12—volume for produced hydrogen.

Since the whole volume occupied by the produced hydrogen is properly measured, we prefer to calculate the quantity of produced hydrogen via the ideal gas equation. For this aim, we measure the pressure, No. 10, and the temperature inside the volume, 12.

As we performed experiments with small quantities of Al chips (ca. 0.6 g) and low alkali concentrations (up to 1M), the reactor warmed up, due to the exothermic reaction, on average by seven to five degrees Celsius. Sacrificing accuracy at the highest reaction rates, for the sake of simplicity, we built an experimental stand without temperature stabilization of the reactor volume. All of the experiments were carried out at a constant laboratory temperature of 20 °C.

4. Results and Discussion

4.1. Experimental Parameters and Controls

4.1.1. Stirring Adjustment at the Surface Reaction Rate Step

Experiments performed with 0.6 g of Al chips in 50 mL of an alkali solution were assumed to run with excess alkaline. At the surface chemical reaction step, sufficient stirring provides a fresh surface contact between solid aluminum and alkali ions. In these cases, the maximum hydrogen production rate, mL/s, is proportional to the alkali molar concentration. In our 100 mL reactor, magnetic stirring worked really efficiently; 300 rpm is enough to achieve a good linearity, as shown in Figure 4. It is worth noting that the determined 300 rpm fits well with [18], describing the mixing of considerably larger aluminum particles by size, in which the difference becomes small between 300 and 400 rpm.

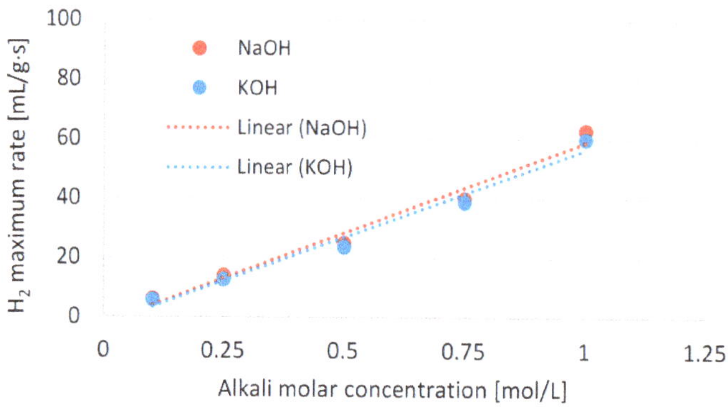

Figure 4. Stirring adjustment by reaching linearity.

4.1.2. Temperature Control during Experiments

For more clarity, as an example, we present temperature fluctuations during the chemical reaction and hydrogen production rate in Figure 5. As the heat release is proportional to the hydrogen production reaction rate, and the peaks for both curves match well, the thermal inertia of the half-filled 100 mL reactor volume is sufficiently low.

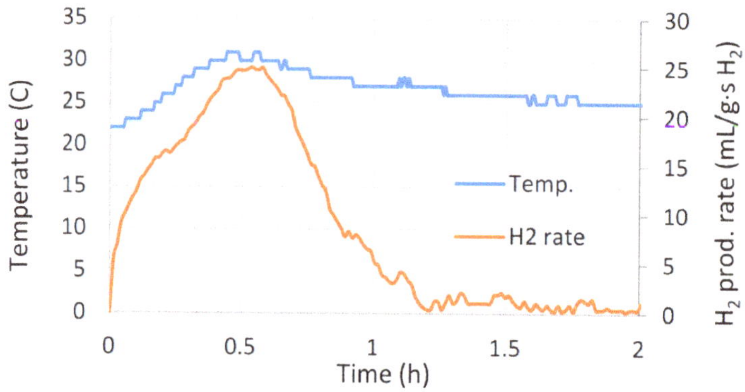

Figure 5. Temperature fluctuations and H_2 production rate during the chemical reaction with 0.5M NaOH.

The measured amplitude of thermal fluctuations, ca. 7 °C and 5 °C before and after the temperature peak, respectively, is rather high for the reaction constant determination. We had previously performed a temperature-dependent series from which the temperature shift of 5 °C with 0.5M alkali caused a difference in the estimated first step rate constant of 18%. Because the paper focuses on comparing results between different alkali concentrations, it is deemed acceptable. Note that the second step rate constant is estimated from the reaction region after the peak, where the amplitude of thermal fluctuations is evidently lower.

4.2. Normalized Cumulative Hydrogen Yield

In Figure 6 we plot the experimentally measured cumulative hydrogen yield, normalized to 1 g of Al, vs. real time. With smaller alkali concentrations of 0.1–0.5M, the yield with NaOH slightly exceeds that with KOH, whereas with 0.75–1M it is the same.

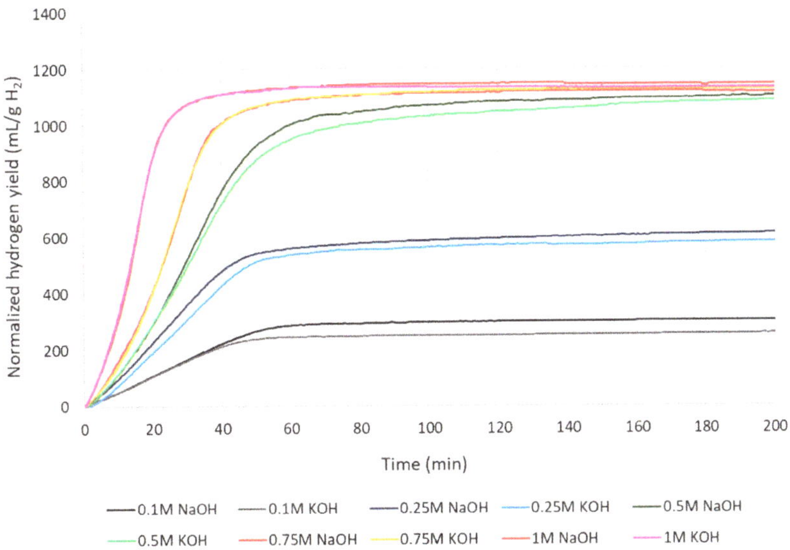

Figure 6. Experimentally measured cumulative H_2 yield.

Similar to the experiments with aluminum flakes in [19], we normalized further our cumulative yield levels to the highest one (1145 mL/g with 1M NaOH) in order to make the comparison, shown in Figure 7. Despite the differences between pure alkalis—NaOH, KOH, and sodium aluminate, $NaAlO_2$—and different temperatures, the trend is similar: after one hour of reaction, the relative hydrogen yield reaches the ca. 95% level at 0.75–0.8M of added aqueous promoter. We consider such alkali concentrations as creating the critical pressure in H_2 bubbles to break the surface film on Al particles in one hour. Nevertheless, temperature dependency is expected in this effect. Our purpose is to investigate this question in our next paper.

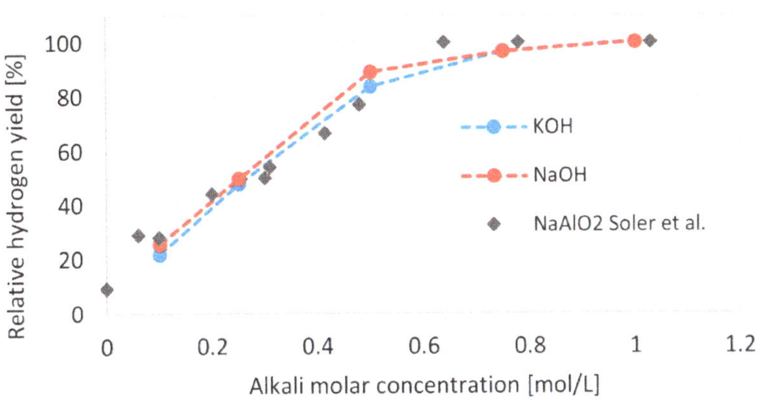

Figure 7. Comparison of normalized cumulative H_2 yields at different alkali concentrations after 1 h of reaction between the present work with NaOH, KOH, and [19].

Many authors perform experiments with alkali concentrations above 1M [17,19–21]. As is known in theory, the maximum H_2 yield of pure aluminum is 1245 mL/g, whereas with our experimental setup we achieved 1145 mL/g with 1M NaOH at moderate stirring. It is 92% of the theoretical maximum, and the deficit of 8% can be mainly attributed to the chemical impurities of the used aluminum. Thus, we consider an alkali concentration up to 1M as appropriate for the profit analysis.

4.3. Hydrogen Production Rate

Figure 8 displays the experimentally measured H_2 production rate. Since H_2 cumulative yields with 1, 0.75M, and 0.5M alkali are very close (Figure 6), the areas below the production rate curves with corresponding alkali concentrations must be very close too. On the other hand, the corresponding three curves expose different slopes (Figure 6), i.e., with a lower alkali concentration the hydrogen production rate is distinctly lower. The mathematical solution to this problem, which was also actually observed in the experiments, is a symmetrical stretching of the production rate curves in the time scale with a lower alkali concentration. Indeed, the maximum hydrogen production rate with 1, 0.75M, and 0.5M alkali is reached at the 16, 27, and 35th minutes, respectively. This notwithstanding, the production rate curves with an alkali concentration less than 0.5M (0.1M and 0.25M) have smaller areas below and less distinct maximums.

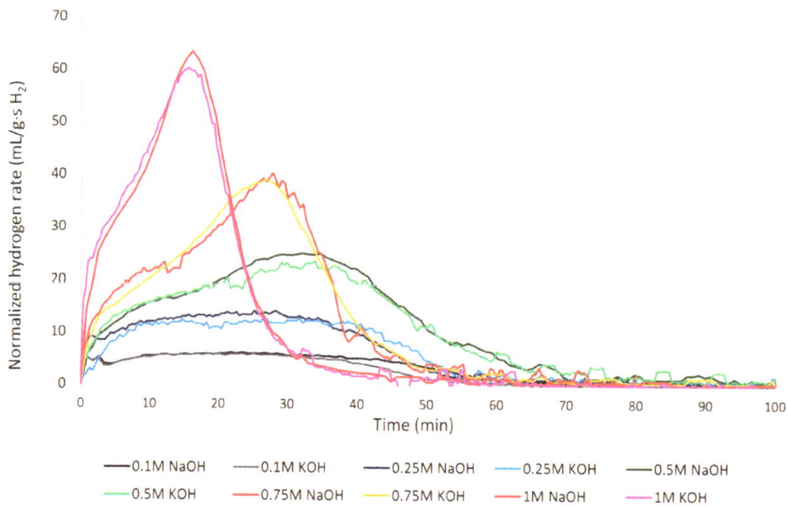

Figure 8. Experimentally measured H_2 production rate.

4.4. Linear Trendline Approximations and Their Compositions

Linear trendline approximations have been carried out with the computer software OriginPro 2016 93E. In order to evaluate the second-order rate constant, i.e., the diffusion coefficient in the byproduct layer, D, in accordance with Equations (15) and (16), we define new particular coordinates.

4.4.1. First-Order Rate Constant for Surface Reaction

Regarding Equations (12) and (13), the best way to evaluate the first-order rate constant, k^*, is to plot the hydrogen yield curves in coordinates: $x = \frac{M_{Al}}{\rho_{Al}} \cdot bc_{alk}t$ and $y = \alpha r_0$. The sloping of the best part of yield growth, measured by the linear trendline method, provides the value of k^* for each experimental condition, depicted in Figure 9. Regarding the lack of temperature stabilization, discussed in Section 4.1.2., we estimate the error of k^* as being around ± 1.5 µm/s from Figure 9, and we assume that the first-order rate constant is independent from the alkali concentration at least up to 1M. This finding coincides with the theoretical opinion that, at the properly performed surface reaction step, the reaction rate, k^*, is linearly proportional to c_{alk} [15].

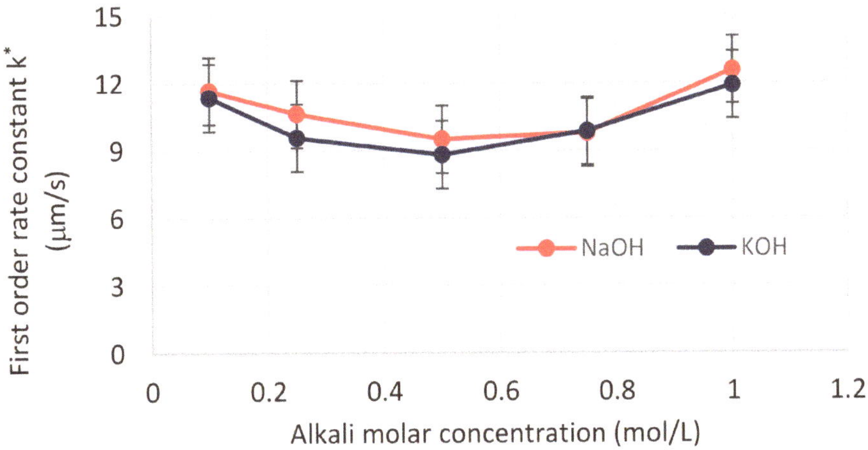

Figure 9. Evaluated first-order rate constants, k^*, for surface reactions at different alkali concentrations.

The authors of [16] reported the values with a powdered aluminum reagent (Kojundo Chemical Laboratory Co., Ltd., Sakado, Japan) as having 99% mass purity and a particle size of 180–425 μm, with 0.5M NaOH of 1.97 μm/s at 20 °C and 22.8 μm/s at 40 °C. Since the average temperature in our reactor, at the maximum reaction rate, was measured to obtain the value of k, Equation (9), with 0.5M NaOH, is ca. 25 °C (Figure 5); the value of k^* of 9.5 μm/s (Figure 9) seems to be consistent. A precise comparison is not possible as the value is strongly influenced by temperature, Al material, and its characteristic size.

Noting the error bars, the difference between NaOH and KOH in Figure 9 is not proven. Nevertheless, we point out that some authors have found NaOH a little more active than KOH. This can be explained by the slightly different mechanisms of catalysis for each alkali. The authors of [17] reported that the activation energies for performed experiments are larger in the presence of KOH over NaOH. The activation energy, E_a, of corrosion is related to the exchange current density, i_c; for slower corrosion E_a is larger and i_c is smaller, and vice versa [22]. This allows for the explanation that, in an open corrosion process, the value of i_c in reactions with KOH is a bit less than that with NaOH.

4.4.2. Diffusion Coefficient of an Aqueous Reactant in the Byproduct Layer

In order to evaluate the second-order rate constant, i.e., the diffusion coefficient in the byproduct layer, D, in accordance with Equations (15) and (16) we defined new particular coordinates: $x = \frac{M_{Al}}{\rho_{Al}} \cdot bc_{alk}t$ and $y = \alpha^2 \frac{r_0^2}{2}$.

Figure 10 displays the evaluated values of the diffusion coefficient of an aqueous reactant in the byproduct layer, D, as the second-order reaction rate constant. Based on the restricted quantitative replicability of our experiments, we estimate the error of D as being around ± 0.5 μm^2/s. The decrease in D by increasing the alkali concentration can be explained by the fact that an alkali of a higher concentration creates a denser surface layer of aluminum hydroxide, which hinders the diffusion of an aqueous reactant. Lower values of D with 0.1M have to be explained otherwise. We assume that, at such a low alkali concentration, not all aluminum oxide is being dissolved from the solid surface. Likewise, in Figure 9, the difference between NaOH and KOH at 0.1M and 0.25M in Figure 10 cannot be proven.

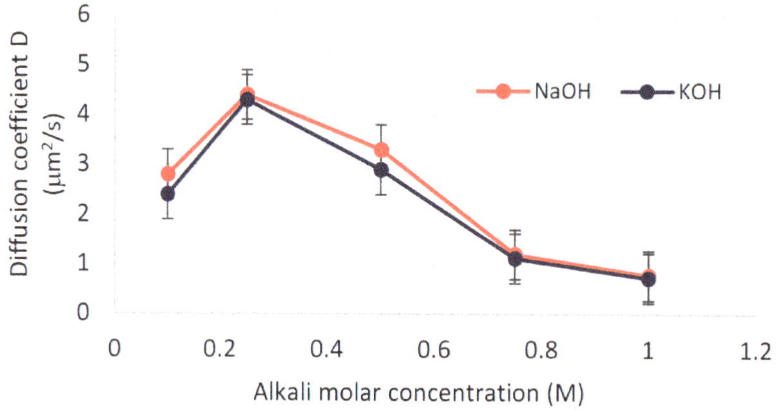

Figure 10. Evaluated second-order rate constants, D, for mass transfer at different alkali concentrations.

4.4.3. Reconstruction of the Hydrogen Yield via Numerically Obtained Values of k^* and D

Figure 11 displays cumulative hydrogen yield curves, reconstrued from the first- and second-order rate parameters, including k^* and D, both obtained through the linear trendline method. In general, the coincidence is quite good. The linear reconstruction shows the induction step of the chemical reaction as a shift on the x-axis. Nevertheless, the accuracy of this parameter in our experiments is under question and is not analyzed in this paper.

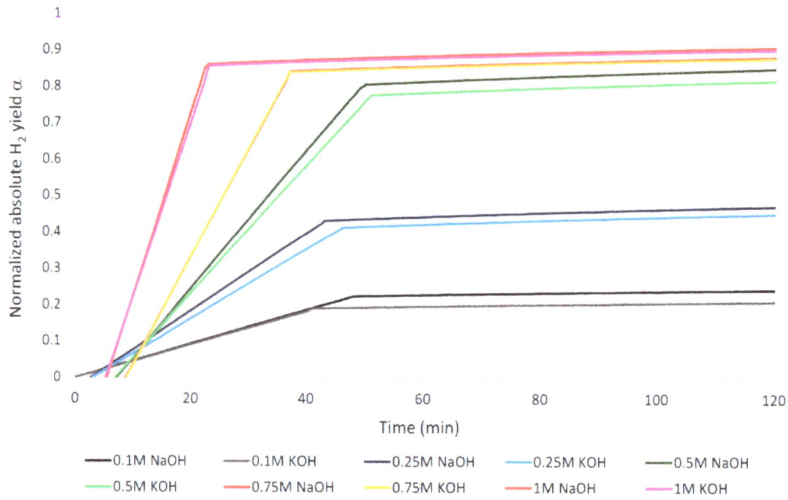

Figure 11. Reconstrued diagrams of cumulative H_2 yield.

4.5. Economic Analysis

4.5.1. Profit Estimation of Performed Experiments

The hydrogen yield curves (Figure 6) and analyses discussed above indicate that, with NaOH or KOH concentrations starting from 0.5M, the yield approaches its real maximum. Hence, in terms of the key criterion of optimizing yield, the conclusion is that any molarity of either NaOH or KOH $\geq$ 0.5M is a potential candidate for the optimal economical production of hydrogen on small and medium scales via hydrolysis.

Considering the optimization of the reaction conditions in terms of economics, the following criterion is the reaction rate. We see in Figure 6 that, at 1M, the reaction completes in about 60 min, compared to 90 min at 0.75M and even up to 160 min at 0.5M. Therefore, in the case of a plant with a target hydrogen production rate, or equivalently a given waste Al throughput rate, the reactor volume must be increased by approximately 1.5 times if an operation opts for production at 0.75M rather than 1M. However, such savings on CAPEX (reactor size) should certainly be considered in the process economics, as it is far from the magnitude that allows clear conclusions.

In our opinion, very few authors who provide a simple profit estimation of the aluminum hydrolysis process pay attention to the loss in cycled alkali. Although NaOH is not consumed by the reaction chain, as shown in Equations (4)–(8), i.e., no NaOH would have to be replaced in an ideal case, elemental analysis indicates that ca. 6–10% of 6M NaOH ends up in a cycled sodium hydroxide solution, even rinsed to a neutral pH after being removed from the reaction volume [23]. There being no detailed investigations on NaOH loss at different initial molarities, it is sensible to assume the relative loss as being proportional to the initial molarity, giving ca. 1.33% with 1M NaOH. Recently, Soberanis et al. [24] and Moreno-Flores et al. [25] published data on the hydrolysis of aluminum cans and waste. The reaction profiles in their pilot reactors broadly agree with our data, and the same conclusions that we make regarding implications for the process economics here are generally backed up by their data and consistent with our findings. In laboratory work, an alkali of a chemical grade is usually used. A good price point of 50 wt.% NaOH solution in water of a chemical grade can be found at Sigma-Aldrich (code 415413), served in 18 L drums. The same NaOH solution of a technical grade, served in 200 L drums (680 lbs.) is seven times cheaper, see Table 2 [26].

Table 2. Profit estimation of aluminum hydrolysis with NaOH promoters of a chemical and technical grade (per 1 kg H_2 yield).

Al Chips 9 kg	**Green H_2 Sale 1 kg**	**Gibbsite Sale 26 kg**	**NaOH Price 26.6 kg 50 wt.% (aq)**	**NaOH Loss of 1.33% (with 1M NaOH)**	**Profit of Al Hydrolysis Process**
EUR 0.5/kg·9 kg = EUR 4.5	EUR 6	EUR 0.2/kg·26 kg = EUR 5.2	Chemical: EUR 547	EUR 7.3	EUR −0.6
			Technical: EUR 76	EUR 1.0	EUR 5.7

Our simple profit estimation is made for NaOH promoters of both a chemical and a technical grade, and is shown in Table 2. In North Europe there is no fixed price for machined waste aluminum chips; we use a value found in reports of EUR 0.5/kg. Regarding the hydrogen yield, there is no sense to distinguish our real yield as 92% of the theoretical (stoichiometric) yield because we take the most frequent price point of European sales, EUR 6–8/kg, which creates an uncertainty of 25% by itself [27]. In this profit estimation, we account for the price of the gibbsite because the expenses of the calcination may depend on the grade of NaOH, which is discussed below. As is seen from Table 2, the difference in profit between applying NaOH of a chemical or technical grade is crucial. If all prices and sales are well estimated, applying 1M NaOH of a chemical grade even leads to negative profit.

There is a dilemma of losing less NaOH vs. increasing the reaction time from 60 min to 90 min by choosing an alkali concentration between 1M and 0.75M. Table 3 deals with calculated "over-profit" by shifting from 1M to 0.75M, related to the overtime of 1 h. As expected, the difference in over-profit between the chemical and technical grades of NaOH is huge, i.e., nine times.

Table 3. Over-profit estimation by shifting from 1M to 0.75M NaOH, related to 1 h of overtime (per 1 kg H_2 yield).

NaOH Price 26.6 kg 50 wt.% (aq)	NaOH Loss of 1.33% (with 1M NaOH)	Profit of Al Hydrolysis Process with 1M	NaOH Loss of 1% (with 0.75M NaOH)	Profit of Al Hydrolysis Process with 0.75M	"Over-Profit" of Decreased NaOH Molarity vs. Overtime
Chemical: EUR 547	EUR 7.3	EUR −0.6	EUR 5.5	EUR 1.2	EUR 3.6/h
Technical: EUR 76	EUR 1.0	EUR 5.7	EUR 0.76	EUR 5.9	EUR 0.4/h

4.5.2. Open Problems of the Further Treatment of Aluminum Hydroxide

To determine the most economical alkali hydroxide concentration for production, or to consider switching from alkali hydroxides to $Ca(OH)_2$ [28,29], two related additional criteria have to be pointed out:

- Hydroxide salt make-up required between runs;
- Impurity levels of chemical elements within the solid aluminum hydroxide $Al(OH)_3$ product and its general suitability for recycling/end use.

The most viable known use of a solid $Al(OH)_3$ product is recycling it back to primary aluminum production. This could constitute "ideal recycling" as opposed to, for example, "down-cycling", in which aluminum is being "sweetened" (diluted) by a significant amount of virgin aluminum from a smelter (20% to 96%, depending on feedstock quality and target alloy) [30–32]. Other examples of "down-cycling" are common plastic recycling and even the current state of the art of most aluminum scrap recycling, where the impurity levels of chemical elements require that they are recycled.

In order to return the solid $Al(OH)_3$ product to aluminum smelters, the hydroxide needs to be converted into oxide or alumina, Al_2O_3. Ideal recycling would be enabled if the hydrolysis product does not incorporate, or entrain, more alkali metal or Ca than the smelter specification accepts. Aluminum smelters typically have accepted specifications on the surface areas (typically ca. 75 m^2/g) and impurity levels of alumina (e.g., $\leq$0.2 wt.% for elements such as Na, K, or Ca). Given an acceptably low impurity profile, aluminum hydroxide can be converted into alumina via calcination either as part of the recycling operation itself or by shipping the solid $Al(OH)_3$ product to an existing bauxite processing facility. In the case of shipping the hydroxide from the hydrolysis recycler, it can be co-introduced with the ore early in the extraction process or directly co-mingled with derived gibbsite, γ-$Al(OH)_3$, via the Bayer process at the final calcination step. Generally, calcination on a large scale at this time is a very inefficient process in terms of energy. The thermodynamic minimum with which to treat 1 kg of Al is only 0.1 kWh, whereas publications report real energy use of 1.6 kWh [33]. In another event, gibbsite, being the dominant crystal phase of the aluminum hydroxide in ore and also the hydroxide precipitate formed within the Bayer process itself [34], would be the preferred phase, since the widely used standard calcination conditions of the hydroxide reliably results in an alumina powder that meets the specific surface area and other specifications for the alumina feedstock of smelters. This is not a trivial consideration, since the formation of the less common synthetic aluminum hydroxide phase bayerite, α-$Al(OH)_3$, and the undesirable incorporation of a sufficient amount of cations, such as Na^+ or Ca^{2+}, to form XRD-detectable amounts of species, such as sodium aluminate or kotoite, as part of the solid hydrolysis product, have been frequently reported under standard hydrolysis conditions [16,27,28]. The formation of bayerite is less of a problem than the challenge posed by impurities. Bayerite often naturally transforms into gibbsite, likely because it is entropically favored, or one could probably research and demonstrate calcination conditions for bayerite not too distinct from those for gibbsite, where the resulting alumina meets the surface area specifications of smelters [34]. However, the impurities will likely play a key role in the morphology and surface area upon calcination; it is known that even small amounts of impurities such as fluoride, chloride, and alkali cations or alkali oxides (that can generally be reduced by careful product washing), or larger amounts of impurities such as carbonate (that is hard to remove fully by washing once resent) can play a critical role in determining the phase and morphology, including the pore structure and surface area, of the alumina powder produced from hydroxide via a given calcination protocol [35]. Additional difficulties in gibbsite calcination to alumina after applying NaOH of a technical grade are not reliably found in the literature.

5. Conclusions

The activation energy for flat Al chips is determined to be 48.1 kJ/mol, which coincides with referred E_a = 40.9 kJ/mol with a 0.02 mm-thick foil and 57.2 kJ/mol with 0.5 thick plates. The cumulative hydrogen yield, achieved in our experiments, reaches 1145 mL per 1 g of Al with 1M NaOH at moderate stirring, which is 92% of the theoretical (stoichiometric) H_2 yield of pure aluminum, 1245 mL/g. The evaluated first-order rate constant of the chemical reaction, k^*, is rather independent from the alkali concentration up to 1M. This notwithstanding, the diffusion coefficient of the aqueous reactant in the byproduct layer, D, decreases four times at increasing alkali concentrations, from 0.25M up to 1M. In Section 4 we provided our reflections on this. For the present type of Al chips, the determined values of k^* and D have sufficient accuracy to be used in a mathematical simulation of chemical and physical processes. Although NaOH shows a bit more activity as a promoter than KOH, noting the error bars, this difference is not proven. The profit estimation of aluminum hydrolysis reveals a crucial difference between applying NaOH of a chemical or technical grade, if attention is paid to the loss in cycled alkali. It seems possible to reduce the expenses due to alkali by applying it at a lower concentration, down to 0.75M, allowing extra time for the reaction to proceed.

Author Contributions: Conceptualization and supervision, J.K., D.M. and C.R.; methodology, P.L., A.K., M.U., S.V. and C.R.; investigation, P.L., A.M., S.V., M.U. and C.R.; yheoretical model and calculations, A.M. and C.R.; writing—original draft preparation, A.M. and C.R.; writing—review and editing, A.K., J.K., D.M., S.V. and C.R.; visualization, A.M., C.R. and M.U.; project administration, A.K.; funding acquisition, S.V., A.K., C.R. and D.M. All authors have read and agreed to the published version of the manuscript.

Funding: This research was funded by the Baltic Research Programme project No. EEA-RESEARCH-92, EEA Grant No. EEZ/BPP/VIAA/2021/5.

Data Availability Statement: Data is contained within the article.

Conflicts of Interest: The authors declare no conflict of interest.

References

1. Calder, G.V.; Stark, T.D. Aluminum reactions and problems in municipal solid waste landfills. *Pract. Period. Hazard. Toxic Radioact. Waste Manag.* **2010**, *14*, 258–265. [CrossRef]
2. International-aluminium.org. Primary Aluminium Production—International Aluminium Institute. Available online: https://international-aluminium.org/statistics/primary-aluminium-production/ (accessed on 29 January 2023).
3. Martínez, S.S.; Benítes, W.L.; Gallegos, A.A.Á.; Sebastián, P.J. Recycling of aluminum to produce green energy. *Sol. Energy Mater. Sol. Cells* **2005**, *88*, 237–243. [CrossRef]
4. Hiraki, T.; Akiyama, T. Exergetic life cycle assessment of new waste aluminium treatment system with co-production of pressurized hydrogen and aluminium hydroxide. *Int. J. Hydrog. Energy* **2009**, *34*, 153–161. [CrossRef]
5. Razavi-Tousi, S.S.; Szpunar, J.A. Mechanism of Corrosion of Activated Aluminum Particles by Hot Water. *Electrochim. Acta* **2014**, *127*, 95–105. [CrossRef]
6. Elsarrag, E.; Elhoweris, A.; Alhorr, Y. The production of hydrogen as an alternative energy carrier from aluminium waste. *Energy Sustain. Soc.* **2017**, *7*, 9. [CrossRef]
7. Xuan, J.; Leung, M.K.H.; Leung, D.Y.C.; Ni, M. A review of biomass-derived fuel processors for fuel cell systems. *Renew. Sustain. Energy Rev.* **2009**, *13*, 1301–1313. [CrossRef]
8. Amendola, S.C.; Sharp-Goldman, S.L.; Janjua, S.M.; Spencer, N.C.; Kelly, M.T.; Petillo, P.J.; Binder, M. A Safe, portable, hydrogen gas generator using aqueous borohydride solution and Ru catalyst. *Int. J. Hydrog. Energy* **2000**, *25*, 969–975. [CrossRef]
9. Wang, E.D.; Shi, P.F.; Du, C.Y.; Wang, X.R. A mini-type hydrogen generator from aluminum for proton exchange membrane fuel cells. *J. Power Sources* **2008**, *181*, 144–148. [CrossRef]

10. Gai, W.Z.; Liu, W.H.; Deng, Z.Y.; Zhou, J.G. Reaction of Al powder with water for hydrogen generation under ambient condition. *Int. J. Hydrog. Energy* **2012**, *37*, 13132–13140. [CrossRef]
11. Bunker, B.C.; Nelson, G.C.; Zavadil, K.R.; Barbour, J.C.; Wall, F.D.; Sullivan, J.P.; Windisch, C.F.; Engelhardt, M.H.; Baer, D.R. Hydration of Passive Oxide Films on Aluminum. *J. Phys. Chem. B* **2002**, *106*, 4705–4713. [CrossRef]
12. Deng, Z.Y.; Ferreira, J.M.F.; Tanaka, Y.; Ye, J. Physicochemical Mechanism for the Continuous Reaction of γ-Al_2O_3-Modified Aluminum Powder with Water. *J. Am. Ceram. Soc.* **2007**, *90*, 1521–1526. [CrossRef]
13. Deng, Z.Y.; Ferreira, J.M.F.; Sakka, Y. Hydrogen-generation materials for portable applications. *J. Am. Ceram. Soc.* **2008**, *91*, 3825–3834. [CrossRef]
14. Omran, M.P.; Mousa Oskueyan, G.; Farhadi, M.; Sedighi, K. Design of Hydrogen Generator for on-Board Hydrogen Generation from Waste Aluminum Chips. In Proceedings of the 22th International Symposium on Transport Phenomena, TUDelft, The Netherlands, 8–11 November 2011; pp. 25–27.
15. Levenspiel, O. *Chemical Reaction Engineering*, 3rd ed.; Wiley: New York, NY, USA, 1998; p. 704.
16. Hiraki, T.; Takeuchi, M.; Hisa, M.; Akiyama, T. Hydrogen production from waste aluminum at different temperatures, with LCA. *Mater. Trans.* **2005**, *46*, 1052–1057. [CrossRef]
17. Porciúncula, C.B.; Marcilio, N.R.; Tessaro, I.C.; Gerchmann, M. Production of hydrogen in the reaction between aluminum and water in the presence of NaOH and KOH. *Braz. J. Chem. Eng.* **2012**, *29*, 337–348. [CrossRef]
18. Ambaryan, G.N.; Vlaskin, M.S.; Dudoladov, A.O.; Meshkov, E.A.; Zhuk, A.Z.; Shkolnikov, E.I. Hydrogen generation by oxidation of coarse aluminum in low content alkali aqueous solution under intensive mixing. *Int. J. Hydrogen Energy* **2016**, *41*, 17216–17224. [CrossRef]
19. Soler, L.; Candela, A.M.; Macanás, J.; Muñoz, M.; Casado, J. In situ generation of hydrogen from water by aluminum corrosion in solutions of sodium aluminate. *J. Power Sources* **2009**, *192*, 21–26. [CrossRef]
20. Soler, L.; Macanás, J.; Muñoz, M.; Casado, J. Aluminum and aluminum alloys as sources of hydrogen for fuel cell applications. *J. Power Sources* **2007**, *169*, 144–149. [CrossRef]
21. Berna, X.S.; Genesca, M.M.; Daga-Monmany, J.M.; Mujal-Rosas, R. Analysis of Valorization Process of Aluminum Breakage Scraps to Obtain Green Hydrogen. *Adv. Valuab. Met. Recycl.* **2021**, *11*, 598.
22. Brett, C.M.A.; Maria, A.N.A.; Brett, O. Principles, Methods, and Applications. *Electrochemistry* **1993**, *67*, 444. Available online: http://www.getcited.org/pub/102998784 (accessed on 29 January 2023).
23. Haller, M.Y.; Amstad, D.; Dudita, M.; Bäuerle, Y. AlEnCycles—Aluminium-Redox-Cycles for the Production of Heat and Electricity for Buildings Based on Renewable Energies. Zenodo. Available online: https://zenodo.org/record/6393255#.ZBjQkMJBy70 (accessed on 20 March 2023).
24. Soberanis, M.A.E.; Vales-Pinzón, C.; Hernández-Núñez, E.; Flota-Bañuelos, M.; Medina, J.; Quintal-Palomo, R.; San-Pedro, L.; Ruiz-Gómez, M. Temperature dependence on hydrogen production from hydrolysis reaction of recycled aluminum. *Clean Techn Environ Policy* **2023**, *25*, 35–49. [CrossRef]
25. Moreno-Flores, R.; Loyola-Morales, F.; Valenzuela, E.; Sebastian, P.J. Design and performance evalu-ation of a prototype hydrogen generator employing hydrolysis of aluminum waste. *Clean Techn Environ Policy* **2023**, *25*, 747–753.
26. Sodium Hydroxide/Caustic Soda Supplier & Distributor | Univar Solutions | Univar Solutions. Available online: https://www.univarsolutions.com/product-categories/essential-chemicals-ingredients/liquid-caustic-soda?hasprice=cspweb (accessed on 20 March 2023).
27. Hydrogen Cost and Sales Prices | H2Valleys. Available online: https://h2v.eu/analysis/statistics/financing/hydrogen-cost-and-sales-prices (accessed on 20 March 2023).
28. Gai, W.Z.; Wang, L.Y.; Lu, M.Y.; Deng, Z.Y. Effect of low concentration hydroxides on Al hydrolysis for hydrogen production. *Energy* **2023**, *268*, 126731. [CrossRef]
29. Kanehira, S.; Kanamori, S.; Nagashima, K.; Saeki, T.; Visbal, H.; Fukui, T.; Hirao, K. Controllable hydrogen release via aluminum powder corrosion in calcium hydroxide solutions. *J. Asian Ceram. Soc.* **2013**, *1*, 296–303. [CrossRef]
30. Zhu, Y.; Cooper, D.R. An Optimal Reverse Material Supply Chain for U.S. Aluminum Scrap. *Procedia CIRP* **2019**, *80*, 677–682. [CrossRef]
31. Soo, V.K.; Peeters, J.R.; Compston, P.; Doolan, M.; Duflou, J.R. Economic and Environmental Evaluation of Aluminium Recycling based on a Belgian Case Study. *Procedia Manuf.* **2019**, *33*, 639–646. [CrossRef]
32. Modaresi, R.; Müller, D.B. The role of automobiles for the future of aluminum recycling. *Env. Sci. Technol.* **2012**, *46*, 8587–8594. [CrossRef] [PubMed]
33. Haller, M.Y.; Amstad, D.; Dudita, M.; Englert, A.; Häberle, A. Combined heat and power production based on renewable aluminium-water reaction. *Renew. Energy* **2021**, *174*, 879–893. [CrossRef]

34. Gale, J.D.; Rohl, A.L.; Milman, V.; Warren, M.C. An ab Initio Study of the Structure and Properties of Aluminum Hydroxide: Gibbsite and Bayerite. *J. Phys. Chem. B* **2001**, *105*, 10236–10242. [CrossRef]
35. Oxides and Hydroxides of Aluminum. WorldCat.org. Available online: https://www.worldcat.org/title/oxides-and-hydroxides-of-aluminum/oclc/18997314 (accessed on 18 March 2023).

energies

MDPI

Article

Stability Studies of Highly Active Cobalt Catalyst for the Ammonia Synthesis Process

Magdalena Zybert [1,*], Hubert Ronduda [1], Wojciech Patkowski [1], Weronika Rybińska [1], Andrzej Ostrowski [1], Kamil Sobczak [2] and Wioletta Raróg-Pilecka [1]

1 Faculty of Chemistry, Warsaw University of Technology, Noakowskiego 3, 00-664 Warsaw, Poland; hubert.ronduda.dokt@pw.edu.pl (H.R.); wojciech.patkowski@pw.edu.pl (W.P.); weronika.rybinska.stud@pw.edu.pl (W.R.); andrzej.ostrowski@pw.edu.pl (A.O.); wioletta.pilecka@pw.edu.pl (W.R.-P.)

2 Biological and Chemical Research Centre, University of Warsaw, Żwirki i Wigury 101, 02-089 Warsaw, Poland; ksobczak@cnbc.uw.edu.pl

* Correspondence: magdalena.zybert@pw.edu.pl

Abstract: Ammonia is currently considered a promising compound for the chemical storage of hydrogen and as an energy carrier. However, large-scale ammonia production is not possible without an active and stable catalyst enabling efficient, long-term work without the need for its replacement. In this paper, the extended stability studies of the highly active promoted cobalt catalyst for ammonia synthesis were carried out. The long-term activity measurements in NH_3 synthesis reaction under conditions close to the industrial ones (400–470 °C, 6.3 MPa, $H_2/N_2 = 3$) were compiled with the characterization of catalyst properties on different stages of its work using N_2 physisorption, XRPD, STEM-EDX, and H_2-TPD. The accelerated aging method was used to simulate the deterioration of catalyst performance during industrial operation. Textural and structural characteristics revealed that the tested catalyst is highly resistant to high temperatures. The lack of significant changes in the specific surface area, morphology of the catalyst particles, surface distribution of elements, and chemisorption properties of cobalt surface during long-term heating (436 h) at 600 °C suggests that stable operation of the catalyst is possible in an ammonia synthesis reactor in the temperature range of 400–470 °C without the risk of losing its beneficial catalytic properties over time. The decline in catalyst activity during the long-term stability test was less than 10%.

Keywords: ammonia synthesis; cobalt catalyst; stability; catalyst deactivation; accelerated aging

1. Introduction

A substantial negative impact on the environment caused by the energy production from fossil fuels makes it necessary to look for new, clean, and effective energy sources. Nowadays, hydrogen is considered the most environmentally friendly fuel [1–3]. Currently, hydrogen is mainly produced from methane by steam reforming with only a small contribution from other more sustainable technologies such as water electrolysis [3–7]. Steam reforming of methane is a mature technology and, at the same time, the least expensive for large-scale hydrogen production, which is mainly used for ammonia production and petroleum refining. These two processes account for about 82% of the total industrial hydrogen demand [8,9]. For other applications, the transport and storage of hydrogen is required, which is an extremely challenging issue due to its specific properties, e.g., a very high energy density by mass, a very low energy density by volume, diffusion through the materials, wide flammability limit ranges with both air and oxygen, and low boiling point (−252.9 °C). Hence, an interesting solution seems to be chemical storage, when hydrogen is bonded in molecules such as methanol, methane, metal amine salts (e.g., $Mg(NH_3)_6Cl_2$), ammonia, or hydrides (e.g., LiH, NaH, KH, $LiBH_4$, $NaBH_4$), which can be easily transported and decomposed on demand, giving hydrogen [10,11]. Among them, one of the most promising compounds is ammonia, containing 17.6% hydrogen by weight, having high energy density per unit volume and acting as a green fuel due to the absence of carbon content, resulting in zero greenhouse gas emissions during combustion. Ammonia also has a significantly higher boiling point (−33.8 °C) than hydrogen, making its storage and transportation relatively much easier than hydrogen [12]. So far, ammonia has been studied as a potential fuel for internal combustion engines, gas turbines, and other industrial purposes [12–15]. Ammonia production technology has been well developed for many years, and consolidated infrastructures for ammonia transport and storage already exist. Nevertheless, the research efforts toward new methods for ammonia synthesis as an alternative to the Haber–Bosch process are being made [16].

Large-scale ammonia production is not possible without the presence of a catalyst. Nowadays, when the efficient utilization of raw materials and energy is of utmost importance, it is desirable to modernize existing processes in order to extend the time of their operation. Hence, the stability of the catalyst, ensuring its long-term operation, is, next to the activity, one of the most crucial parameters characterizing the catalyst. Long-term, stable operation of the catalyst, without the need for its replacement, reducing the cost of plant shutdown, is the key factor qualifying it for industrial use. Numerous chemical and physical factors influence catalyst stability, resulting in deactivation [17]. Deactivation of the catalyst is inevitable, but its time varies significantly for the different processes (e.g., seconds for the cracking catalyst, months or a year for the hydrodesulfurization catalyst, or even 15 years for the iron ammonia synthesis catalyst). Deactivation can manifest itself mostly by changes in activity and selectivity over time. The most important causes of the catalyst deactivation include (1) poisoning, (2) fouling, (3) thermal degradation, (4) mechanical deactivation, and (5) corrosion/leaching [18–22]. Some of these mechanisms are chemical in nature and some are mechanical. However, it should be remembered that deactivation is a complex phenomenon, and the causes are often not independent. Moreover, it can occur at different levels, i.e., for specific active sites, catalyst particles, or even for the whole catalytic bed. Some deactivation mechanisms are reversible, and catalysts can often be regenerated before they ultimately have to be replaced. In some cases, damage is permanent and must be prevented or mitigated at least to maintain the catalyst lifetime, which is of crucial importance for the economics of a process. A catalyst deactivation generates costs connected not only with the replacement of catalyst charge but also with increasing raw materials and energy usage, cost of plant shutdown, lost production, and environmental costs of waste catalyst disposal. Studies of the catalyst's stability are required to predict the expected catalyst's lifetime in the plant and its optimization [23].

Although the definition of catalyst suggests that a catalyst can be used forever, it cannot be true in real industrial conditions due to numerous chemical and physical factors gradually decreasing activity/selectivity with the running time up to a certain value, qualifying it as inactive. In general, catalyst deactivation can be investigated by measuring activity as a function of time. However, some techniques are applicable only for industrial research because it is difficult to create conditions close to the industrial ones in a lab-scale study. This applies primarily to processes carried out in severe temperature and pressure conditions, and the catalyst's lifetime is strictly dependent on these conditions. In addition, reliable validation of catalyst stability requires monitoring activity changes over a long period of time. For academic research, the accelerated aging method based on applying high temperatures and/or high concentrations of poisons is useful to simulate the deterioration of catalyst performance under industrial conditions and achieve similar catalyst deactivation in a substantially shorter time [24]. The reliability of this method strongly depends on test conditions, which should differ as little as possible from those of the industrial reactor, except for one parameter chosen to give the acceleration (e.g., temperature) [25]. This methodology is commonly applied for thermal stability tests of most heterogeneous catalysts.

The iron catalyst for ammonia synthesis is a good representative of a catalyst in which activity is affected by thermal deactivation (slow sintering) when working in modern plants, where feed gas purification is very effective, thus minimizing the effects of poisoning. For example, stability studies for this catalyst were conducted by Liu et al. [26], who subjected the commercial iron catalyst (A301) based on wustite to a heat-resistance stability test at 500 °C for 20 h (5 MPa, 30,000 h^{-1}), demonstrating the high stability of this catalyst. The loss of the ammonia concentration in the outlet gas was not higher than 0.5% and comparable to other iron catalysts based on magnetite. A similar procedure for wustite-based and magnetite-based catalysts was used by Pernicone et al. [27]. After the treatment at 600 °C for 16 h in synthesis gas (3 MPa, 20,000 h^{-1}), both studied catalysts retained about 80% of their initial activity. In the case of Ru catalysts, Lin et al. [28] performed the stability test of carbon-supported Ru catalysts by a high-temperature treatment process at 500 °C for 40 h. The results indicated a significant decrease in the ammonia synthesis rates for K-Ru/C and Ba-K-Ru/C by about 37% and 29%, respectively, compared to the initial rate. In the paper of Kowalczyk et al. [29], the short-term overheating of carbon-supported Ru catalysts at 520 °C for 24 h in $3H_2 + N_2$ mixture was carried out. A threefold decrease in activity was observed for the K + Ru/C catalyst. However, the Ba + Ru/C catalyst was revealed to be strongly resistant to overheating—only a several percent decrease in the reaction rate was observed compared to the initial rate.

In our previous published papers [30–32], a detailed investigation was conducted to evaluate the catalytic performance and characteristics of the cobalt catalyst promoted with cerium and barium (CoCeBa). The studies revealed that the catalyst is highly active in ammonia synthesis under conditions close to the industrial ones. The activity of the CoCeBa catalyst was higher than that of the commercial iron catalyst (KM I, H. Topsoe) [30,31]. Moreover, the kinetic characteristics of the catalyst showed that it is much less sensitive to the ammonia content in the gas, especially at high ammonia content [31], than the iron catalyst. The present study is the continuation of our systematic studies on the CoCeBa catalyst and investigates the stability of this catalyst. The long-term activity in ammonia synthesis was analyzed with additional measurements characterizing the most important properties of the catalyst and their changes during long-term operation to understand better the changes occurring in this catalyst structure and the adsorption properties of its surface. Three samples of the CoCeBa catalyst were tested, in which the promoter (Ba) was introduced using three different preparative methods: deposition–precipitation, wet impregnation, and wet mixing. An additional aim of the work was to determine the influence of the promoter addition method on the properties and stability of the CoCeBa catalyst.

2. Materials and Methods

2.1. Catalyst Preparation

The mixture of cobalt carbonate and cerium carbonate in a fixed proportion was obtained by co-precipitation method using cobalt(II) nitrate hexahydrate and cerium(III) nitrate hexahydrate as Co and Ce sources and potassium carbonate as a precipitating agent. All the reagents used were of analytical grade. After drying at 120 °C and calcination at 500 °C of the precipitate, the mixture of Co_3O_4 and CeO_2 was obtained. A detailed description of the preparation method used can be found in our previous paper [31]. The obtained oxide mixture was then subjected to further procedures aiming to introduce the Ba promoter using three different preparative methods:

(1) Wet impregnation (WI)—the material was impregnated with an appropriate amount of an aqueous solution of barium(II) nitrate. The solvent (water) was removed using a vacuum evaporator. Then, the sample was dried in air at 120 °C for 18 h.
(2) Wet mixing (WM)—the material was mechanically mixed in a mortar with an appropriate amount of an aqueous solution of barium(II) nitrate and dried in air at 120 °C for 18 h.
(3) Deposition–precipitation (DP)—the material was dispersed in an appropriate amount of an aqueous solution of barium(II) nitrate. Then, an aqueous solution of potassium carbonate was introduced dropwise and, maintaining a constant temperature (90 °C)

and stirring speed (350 rpm), the process of precipitation of barium carbonate on particles of a mixture of cobalt and cerium oxides was carried out. The precipitate was then filtered, washed, and dried at 120 °C for 18 h.

Finally, the samples were crushed and sieved to obtain grains of 0.20–0.63 mm in size. The obtained catalyst precursor samples were denoted as WI, WM, and DP, indicating the method of Ba promoter addition. The chemical composition of the prepared samples is indicated in Table 1. The method of Ba promoter addition caused the differences in textural properties and phase composition of the obtained catalyst precursors. A description of these properties of the catalyst precursors can be found in Table S1, Figures S1–S3.

Table 1. Physicochemical properties of the studied catalysts.

Catalyst	Element Content (wt%) [1]			SSA_{red} ($m^2\ g^{-1}$) [2]	
	Co	Ce	Ba	After Reduction in H_2 Flow at 550 °C for 10 h	After Overheating in H_2 Flow at 600 °C for 10/20/30 h
WI	70.1	8.4	12.6	5.6	3.9/4.2/3.7
WM	70.3	8.2	12.8	13.3	7.2/8.0/6.0
DP	70.7	8.1	12.3	11.2	11.8/12.3/10.4

[1] Element content determined using the XRF method. [2] SSA_{red}—specific surface area determined for the reduced catalysts after heat treatment in H_2 flow at 550 or 600 °C based on the BET adsorption model.

2.2. Catalyst Characterization Methods

N_2 physisorption at −196 °C using an ASAP2020 instrument (Micromeritics Instrument Co., Norcross, GA, USA) was applied to determine the specific surface area (SSA) of the catalyst precursors and the reduced form of the catalysts (SSA_{red}). Prior to the measurements, each precursor sample (500 mg) was subjected to a two-step degassing procedure in a vacuum: at 90 °C for 1 h and then at 150 °C for 4 h. Before the measurement of the catalysts in reduced form, the catalyst sample was firstly reduced in situ at 550 °C for 10 h in H_2 flow and then subjected to degassing at 150 °C for 2 h. After the measurement, the same sample was reduced in situ at 600 °C for 10 h, degassed at 150 °C for 2 h, and the measurement was carried out. The same procedure was repeated after the further in situ reduction at 600 °C for 20 and 30 h. The reduction and degassing were conducted in the apparatus directly before N_2 physisorption measurements. The obtained data were approximated using the Brunauer–Emmett–Teller (BET) isotherm model in the relative pressure range (P/P_0) from 0.05 to 0.3, and the specific surface area (SSA and SSA_{red}) values were determined.

Powder X-ray diffraction patterns were recorded using a Bruker D8 Advance diffractometer (Bruker AXS, Karlsruhe, Germany) equipped with an LYNXEYE position-sensitive detector, using Cu-Kα radiation (λ = 0.15418 nm). The data were collected in the Bragg–Brentano (θ/θ) horizontal geometry for the range 2θ = 10–80° in a continuous scan using 0.03° steps of 10 s/step. Data were collected under standard laboratory conditions.

TEM investigations were performed on a Talos F200X (FEI Company, Hillsboro, OR, USA) microscope operated at 200 kV. The measurements were performed in TEM and scanning TEM (STEM) modes using a high-angle annular dark-field (HAADF) detector and energy-dispersive X-ray spectroscopy on a Brucker BD4 spectrometer. The catalyst samples were investigated after ex situ reduction at 550 °C for 18 h under H_2 flow and after ex situ reduction at 600 °C for 96 h under H_2 flow.

H_2 temperature-programmed desorption using an AutoChem II 2920 instrument (Micromeritics Instrument Co., Norcross, GA, USA) was applied to determine the sorption properties of the catalyst surface. Before the analysis, the precursor sample (500 mg) was reduced in situ at 550 °C for 18 h under H_2 flow and purged at 570 °C for 2 h under Ar flow. Next, the system was cooled to 150 °C, and H_2 was introduced for 15 min. Then, the reactor was cooled to 0 °C and kept at that temperature for 15 min. Next, the sample was purged with Ar at 0 °C until the baseline was stable. Desorption of hydrogen was carried out under Ar flow with a heating rate of 5 °C min^{-1} up to 600 °C. A TCD detector was used to collect the data continuously during the process. The same experiments according to the procedure described above (i.e., hydrogen sorption and subsequent desorption) were conducted after reduction of the catalyst sample under H_2 flow at 600 °C for 18, 30, 66, 84, and 96 h.

2.3. Catalyst Activity and Stability Tests

The performance of the studied catalysts was tested in the NH_3 synthesis reaction. A high-pressure apparatus with a tubular flow reactor was used. Before the measurements, each catalyst sample (500 mg, grain size 0.2–0.63 mm) was subjected to an activation (reduction) procedure in a flow of a stoichiometric H_2/N_2 mixture (gas purity 99.99995 vol%., gas flow rate 30 dm^3 h^{-1}) under the pressure of 0.1 MPa according to the following temperature program: 470 °C for 72 h → 520 °C for 24 h → 550 °C for 48 h. Next, the measurements of ammonia synthesis reaction rate over the catalyst were conducted under steady-state conditions, i.e., 6.3 MPa, 400–470 °C, H_2/N_2 = 3, gas flow rate 70 dm^3 h^{-1}. NH_3 concentration in the outlet gas was determined using an interferometer, and NH_3 synthesis reaction rate was calculated. A detailed description of the experimental setup and calculating methods can be found in [33]. To investigate the stability of the catalyst, the overheating procedure was applied. The catalysts' heat treatment conditions were as follows: 600 °C, 0.1 MPa, H_2/N_2 = 3, gas flow rate 30 dm^3 h^{-1}. The long-term heat treatment was carried out for a total time of 436 h. Catalyst activity measurements were conducted under steady-state conditions, i.e., 6.3 MPa, 400 °C, H_2/N_2 = 3, gas flow rate 70 dm^3 h^{-1} after 16, 82, 98, 266, and 436 h of overheating.

3. Results

3.1. Evaluation of Catalyst Stability in Ammonia Synthesis Reaction

The performance of CoCeBa catalysts in ammonia synthesis was systematically examined as a function of time. All the studied catalysts displayed high initial activity (after standard reduction at 550 °C) in the ammonia synthesis reaction, presented as unfilled marks in Figure 1a. The promoter (Ba) addition method appears to have a slight effect on the activity of the catalysts—the difference in reaction rates resulting from various promoter addition methods ranged from 10 to 25%. The CoCeBa catalyst obtained using the deposition–precipitation method (DP sample) exhibited the highest activity among the studied catalytic systems. Since the accelerated aging method is most appropriate to evaluate the stability of the catalyst on the laboratory scale, a high-temperature treatment at 600 °C for a total time of 436 h was carried out to simulate the deterioration of catalyst performance under industrial conditions. As shown in Figure 1a, the catalytic activities initially increased by approximately 10% for all the catalysts after the first 16 h of treatment at 600 °C compared to the activity after activation. This may be related to the progressive reduction process of the catalysts, connected with the production of an additional promoter ($BaCeO_3$) with strong electron-donating properties, favoring the high activity of the tested catalysts in the ammonia synthesis reaction. The influence of this promoter on CoCeBa catalyst activity and its formation during catalyst reduction was described in detail in our previous paper [32]. The further heating of the catalysts at 600 °C for the next 66 h caused the activity of the DP and WI catalysts to decrease slightly (by 8 and 6%, respectively) or remained unchanged for the WM catalyst. The long-term stability test was continued for a total time of 436 h (over 18 days), and no further changes in the activity of CoCeBa catalysts were observed during their operation. The obtained results evidence that the CoCeBa catalyst is strongly resistant to overheating. This allows us to assume that this type of catalyst can operate stably and effectively even at temperatures higher than the standard operating temperature of ammonia synthesis reactors. Figure 1b presents the temperature dependence of the ammonia synthesis rate of the CoCeBa catalysts at the end of the long-term stability test. The studied catalysts exhibited similar activity across the examined temperature range of 400–470 °C. The apparent activation energies determined from the slopes of the Arrhenius plots (Figure S4) were similar and in the range of 47–49 kJ mol^{-1}.

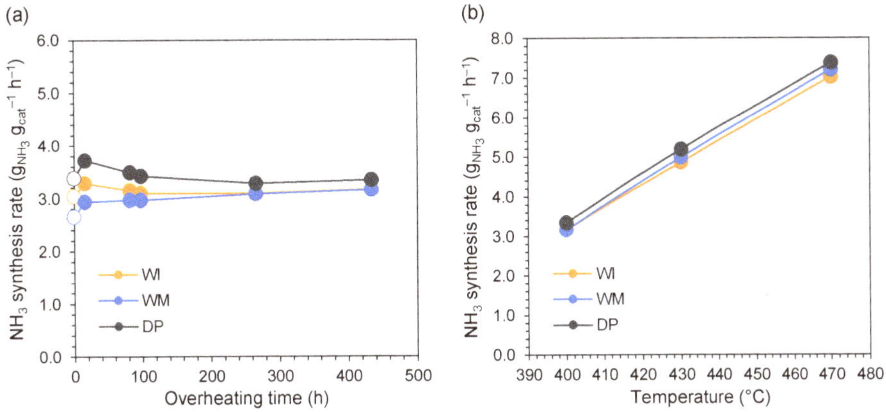

Figure 1. (**a**) Ammonia synthesis rate as a function of overheating time (reaction conditions: 400 °C, 6.3 MPa, H_2/N_2 = 3, 70 dm^3 h^{-1}; heat treatment conditions: 600 °C, 0.1 MPa, H_2/N_2 = 3, 30 dm^3 h^{-1}). Unfilled marks—activity after standard activation (reduction) at a temperature of 550 °C, filled marks—activity after heat treatment at a temperature of 600 °C. (**b**) Temperature dependence of the ammonia synthesis rate at 6.3 MPa, H_2/N_2 = 3, 70 dm^3 h^{-1} after long-term heat treatment at 600 °C for 436 h.

The superior performance of the CoCeBa catalysts for ammonia synthesis was revealed by comparison to the literature results with similar reaction conditions (Table S2) [34–40]. In particular, the ammonia formation rate over CoCeBa catalysts reached 3.17–3.35 g_{NH3} g_{cat}^{-1} h^{-1}, which was almost three times that of commercial iron catalyst (ZA-5). In addition, the CoCeBa catalysts showed much higher activity than similar composite catalysts (Table S2).

3.2. Evaluation of Catalyst Properties during Stability Studies

To explain the effect of the high stability of the CoCeBa catalyst during the long-term stability test, detailed characterization studies were performed focusing on changes in structural and sorption properties of the catalyst surface during the long-term operation.

Firstly, N_2 physisorption analysis of the specific surface area of the catalysts (SSA_{red}) was carried out during long-term heat treatment in the H_2 flow (in situ measurements). As shown in Table 1, the initial values of the SSA_{red} parameter after reduction in the hydrogen flow at a temperature of 550 °C for 10 h were in the range of 5–13 m^2 g^{-1}, depending on the catalyst preparation method. As a result of overheating of the catalysts in the hydrogen flow at a temperature of 600 °C for 10 h, a decrease in the specific surface area was observed by 30 and 45% for the WI and WM samples, respectively. For the DP catalyst, no change in the SSA_{red} value was observed. Further overheating under these conditions for 20 and 30 h did not cause any further changes in the specific surface area of the studied catalysts. This indicates a high resistance to the sintering of the studied catalysts.

Next, the structure of the CoCeBa catalysts at various stages of their operation was investigated using XRPD (ex situ measurements). The main phases identified in the tested catalysts are marked in Figure 2. It is clearly visible that regardless of the composition of the catalyst precursors, resulting directly from the method used to introduce the Ba promoter (Figure S3), the catalysts in the active form, i.e., after reduction in hydrogen flow at 550 °C for 18 h, showed a similar phase composition (Figure 2). For all catalysts, the presence of two metallic cobalt phases (hexagonal and cubic) was observed. However, for catalysts obtained using impregnation methods (WM, WI samples), the reflections of two crystalline forms of cobalt (hexagonal Co and cubic Co) were observed. Meanwhile, for the catalyst obtained using the deposition–precipitation method (DP sample), only reflections from the cubic Co were observed. Moreover, barium cerate ($BaCeO_3$) phase was identified for all the studied catalysts. This phase formed in situ under the conditions of catalyst reduction as a result of the interaction of components containing Ce and Ba. A detailed investigation of the in situ formation of $BaCeO_3$ promoter during the studied catalyst reduction can be found in our previous paper [32].

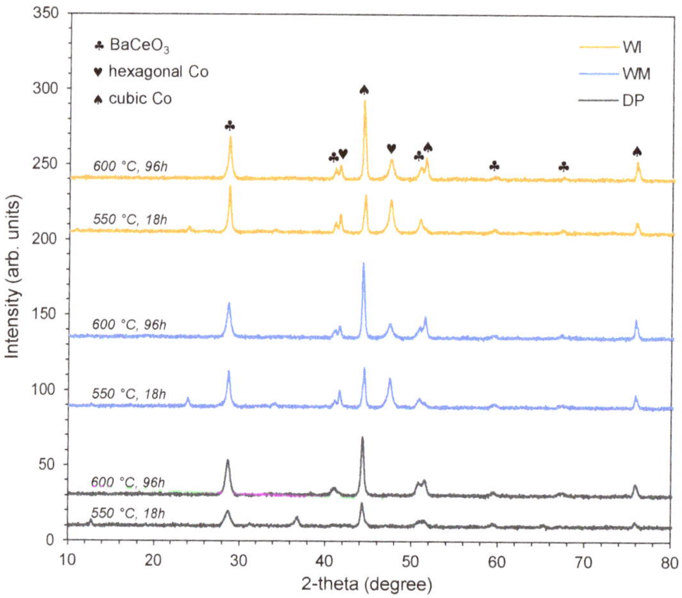

Figure 2. Diffractograms of the CoCeBa catalysts after reduction in hydrogen flow at 550 °C for 18 h and after overheating in hydrogen flow at 600 °C for 96 h.

After overheating the catalysts in hydrogen flow at a temperature of 600 °C for 96 h, no significant changes in the phase composition of the catalysts were observed (Figure 2). The main components of the catalyst were still cubic Co, hexagonal Co, and $BaCeO_3$. However, only reflections from the cubic Co were observed for the catalyst obtained using the deposition–precipitation method (DP sample). In turn, for the catalysts obtained using impregnation methods (WI, WM samples), the intensity of reflections derived from hexagonal Co decreased while the intensity of reflections from cubic Co increased. This may be related to the phase transformation of hexagonal cobalt (hcp) into cubic cobalt (fcc) during high-temperature heating of the catalyst [41–43]. However, changes in the amount of hexagonal Co and cubic Co phases did not significantly influence the size of the Co crystallites (Table 2). After reduction at 550 °C for 18 h, the crystallites of the active phase had a size in the range of 15–17 nm and 20–26 nm for the hexagonal Co and cubic Co phases, respectively. After overheating at a temperature of 600 °C for 96 h, the Co crystallite sizes were in the range of 15–17 nm and 23–27 nm for the hexagonal Co and cubic Co phases, respectively. Moreover, after overheating the catalysts at 600 °C, it was observed that the reflections from the $BaCeO_3$ promoter phase became more intense, indicating increasing crystallinity of this phase. The average $BaCeO_3$ crystallite sizes in the CoCeBa catalysts after overheating at 600 °C were larger than those after heating at 550 °C (Table S2). At the same time, there were no significant differences in the unit cell parameters of the $BaCeO_3$ phases in the CoCeBa catalysts before and after overheating.

Table 2. Average Co crystallite size calculated based on the FWHM value obtained by Rietveld refinement method.

Catalyst	After Reduction in H_2 Flow at 550 °C for 18 h		After Overheating in H_2 Flow at 600 °C for 96 h	
	Hexagonal Co (nm)	Cubic Co (nm)	Hexagonal Co (nm)	Cubic Co (nm)
WI	15	26	17	27
WM	17	25	15	26
DP	–	20	–	23

Small reflections, visible on the diffraction profiles after reduction at 550 °C in the range of the 2-theta = 11–36°, disappeared. They were most likely related to the presence of barium carbonate, which, during long-term overheating in hydrogen, reacted with the second promoter (cerium oxide), resulting in the $BaCeO_3$ phase (in situ formation of the promoter phase) [32].

Microscopic investigations were performed for the catalysts at various stages of the long-term stability test (ex situ measurements) to analyze the morphology and chemical composition of the surface layers of the prepared catalysts. The representative TEM images of the CoCeBa catalysts after reduction in hydrogen flow at 550 °C for 18 h presented in Figure 3 clearly indicate the difference in their morphology depending on the method of the promoter addition. The catalysts obtained using wet impregnation (Figure 3a) and the wet mixing method (Figure 3c) consist of agglomerates with irregular shapes containing grains of different sizes but smaller than 100 nm and 200 nm, respectively. The morphology of the catalyst obtained using the deposition–precipitation method (DP) seems to be different. Figure 3b indicates the significant homogeneity of this catalyst. The DP catalyst particles are characterized by regular shapes of small sizes below 50 nm.

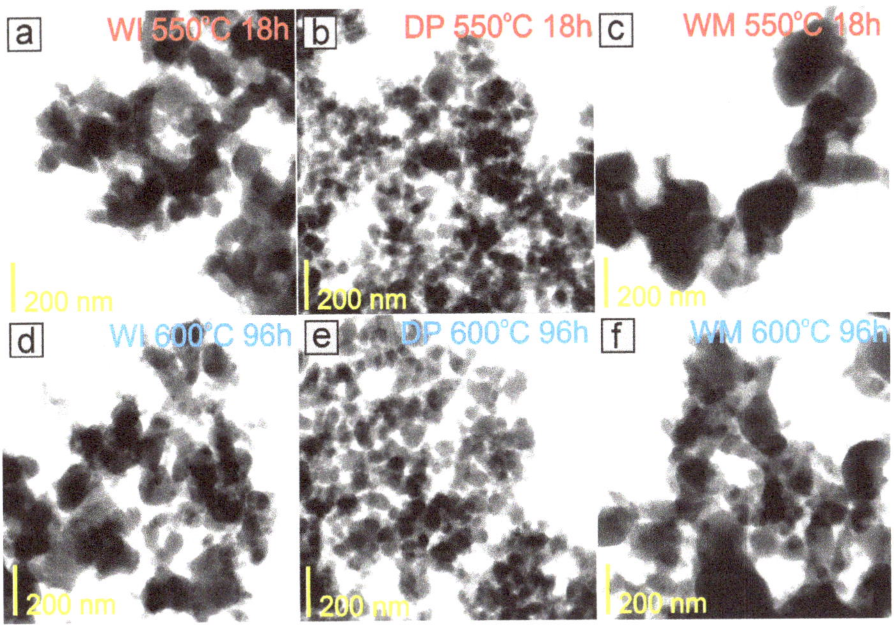

Figure 3. TEM images of the studied WI (**a**,**d**), DP (**b**,**e**), and WM (**c**,**f**) catalysts after reduction in hydrogen flow at 550 °C for 18 h and after overheating in H_2 flow at 600 °C for 96 h (ex situ measurements).

The morphology does not change after overheating the catalysts in H_2 flow at 600 °C for 96 h (Figure 3d–f). This lack of macroscopic changes in the catalysts' morphology proves that the particles are highly resistant to overheating and that there are no sintering effects that could negatively affect the long-term performance of catalysts in the ammonia synthesis reactor. STEM-EDX analysis provided additional information on the chemical composition of the catalysts' surface. It is clearly visible in the EDX maps presented in Figures 4a, 5a and 6a that after reduction in hydrogen flow at 550 °C for 18 h, the homogeneous distribution of Co throughout the catalysts particles was observed for all the studied samples regardless of the promoter addition method used. The distribution maps for cerium, barium, and oxygen indicate that the $BaCeO_3$ promoter is evenly distributed throughout the samples. It does not form separate grains but is relatively well distributed on the surface of the cobalt particles. Only a few areas of greater promoter concentration can be found on the studied catalyst surfaces. After overheating the catalysts in H_2 flow at 600 °C for 96 h, the distribution of the catalyst components remains rather unchanged (Figures 4b, 5b and 6b). This favorable element distribution may promote effective electron donation from the promoter to the active phase, ensuring its efficient catalytic performance during long-term operation [44].

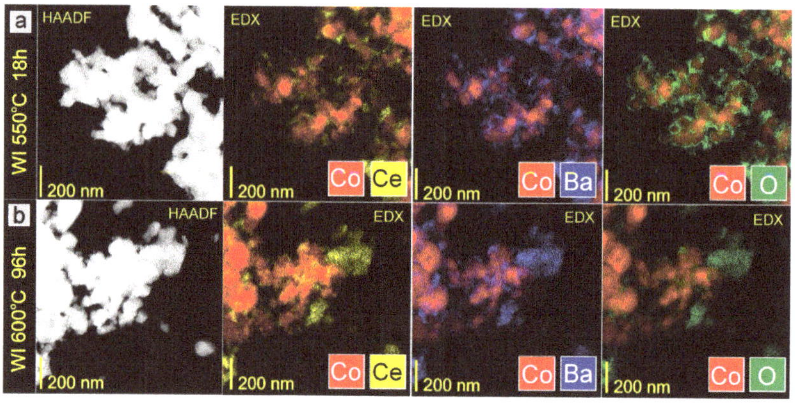

Figure 4. STEM-HAADF images and corresponding EDX distribution maps of elements for WI catalyst: (**a**) after reduction in hydrogen flow at 550 °C for 18 h, (**b**) after overheating in hydrogen flow at 600 °C for 96 h (ex situ measurements).

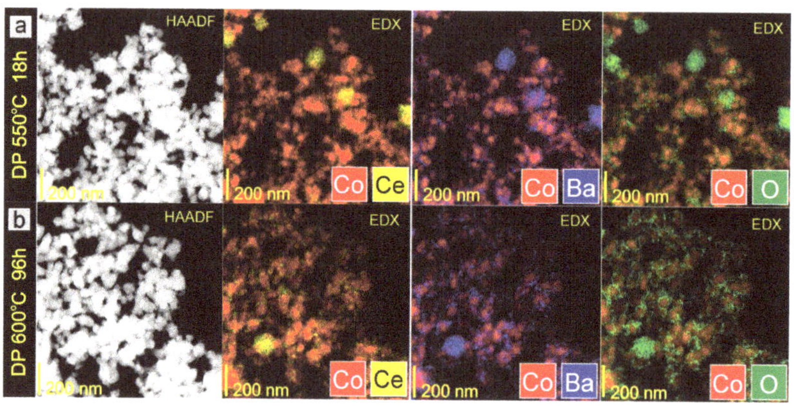

Figure 5. STEM-HAADF images and corresponding EDX distribution maps of elements for DP catalyst: (**a**) after reduction in hydrogen flow at 550 °C for 18 h, (**b**) after overheating in hydrogen flow at 600 °C for 96 h (ex situ measurements).

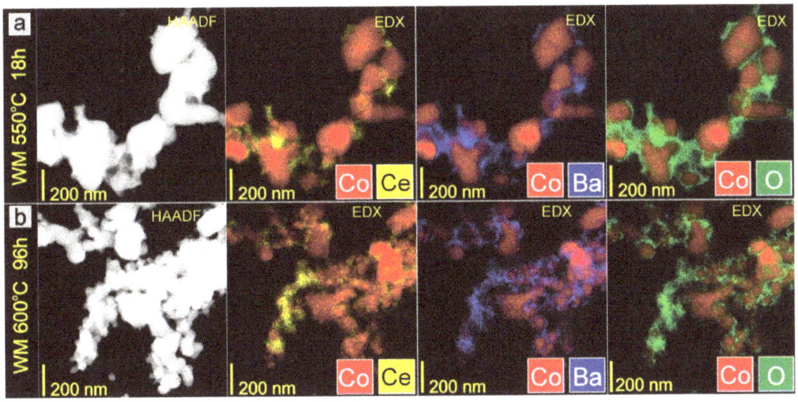

Figure 6. STEM-HAADF images and corresponding EDX distribution maps of elements for WM catalyst: (**a**) after reduction in hydrogen flow at 550 °C for 18 h, (**b**) after overheating in hydrogen flow at 600 °C for 96 h (ex situ measurements).

The adsorption properties of the catalysts' surface were examined using H_2-TPD experiments. Although nitrogen dissociation is considered the rate-determining step of the ammonia synthesis reaction [45–49], the strong chemisorption of the second reactant, i.e., hydrogen, on the catalyst surface may also be an essential factor determining the operation of the catalyst. In particular, strong chemisorption of hydrogen on the catalyst surface, known as the "hydrogen poisoning" phenomenon, may cause the catalyst's low efficiency or rapid performance degradation during long-term operation. Hence, the obtained results in the form of hydrogen desorption curves from the surfaces of the studied catalysts (Figure 7a–c) allowed for the interpretation of the strength of active sites binding hydrogen to the catalyst surface, which is of pivotal importance from the point of view of the mechanism of NH_3 synthesis reaction.

Figure 7a shows the H_2-TPD profiles of the catalyst obtained by wet impregnation (WI sample). After reduction in hydrogen flow at 550 °C for 18 h, two desorption peaks are observed for the WI catalyst in the temperature range of 0–600 °C. The low-temperature (<200 °C) and high-temperature (400–550 °C) desorption peaks can be ascribed to the cobalt active sites that weakly and strongly bind hydrogen, respectively [50,51]. The area ratio of low-to-high-temperature peaks is about 0.6, which means that the sites binding hydrogen strongly to the catalyst surface dominate in the WI catalyst. However, hydrogen desorption from strongly binding sites occurs in a temperature range similar to that in which the ammonia synthesis reaction is carried out. It can, therefore, be concluded that all adsorption sites identified for the WI catalyst are available to the reagents under the reaction conditions [50]. This means the catalyst is free of hydrogen poisoning under the tested conditions. After overheating in hydrogen flow at 600 °C for 18 h, the desorption profile did not change, i.e., the nature of the adsorption sites was preserved. Only the intensity of the low-temperature peak decreased slightly, i.e., the ratio of low-to-high-temperature peaks decreased to 0.4. It should be underlined that no further changes in the hydrogen desorption profile from the surface of this catalyst were observed during further overheating of the catalyst for 30, 66, 84, and 96 h. Similar hydrogen desorption profiles (Figure 7b) were observed for the catalyst obtained using the wet mixing method (WM sample) both after reduction in hydrogen flow at 550 °C for 18 h and after overheating at 600 °C. The temperature ranges for hydrogen desorption for each type of active site were similar to that for the WI catalyst. Only the low-to-high-temperature peak ratio was slightly higher and reached ca. 1.0 after reduction at 550 °C and 0.7 after overheating at 600 °C. Therefore, it is clear that in the case of the WM catalyst, similarly to the WI catalyst, the active sites strongly binding hydrogen on the catalyst surface play a dominant role. A different distribution of adsorption sites was observed for the catalyst obtained using the deposition–precipitation method (DP sample) (Figure 7c). After reduction at 550 °C, the desorption profile consists of a significant low-temperature peak in the range of 0–300 °C and a broad high-temperature peak of very low intensity in the temperature range of 400–600 °C. This indicates the dominance of weakly hydrogen-binding sites on the surface of the DP catalyst. After overheating at 600 °C for 18 h, the high-temperature peak disappeared completely, and the desorption profile did not change further during overheating at 600 °C for 30, 66, 84, and 96 h.

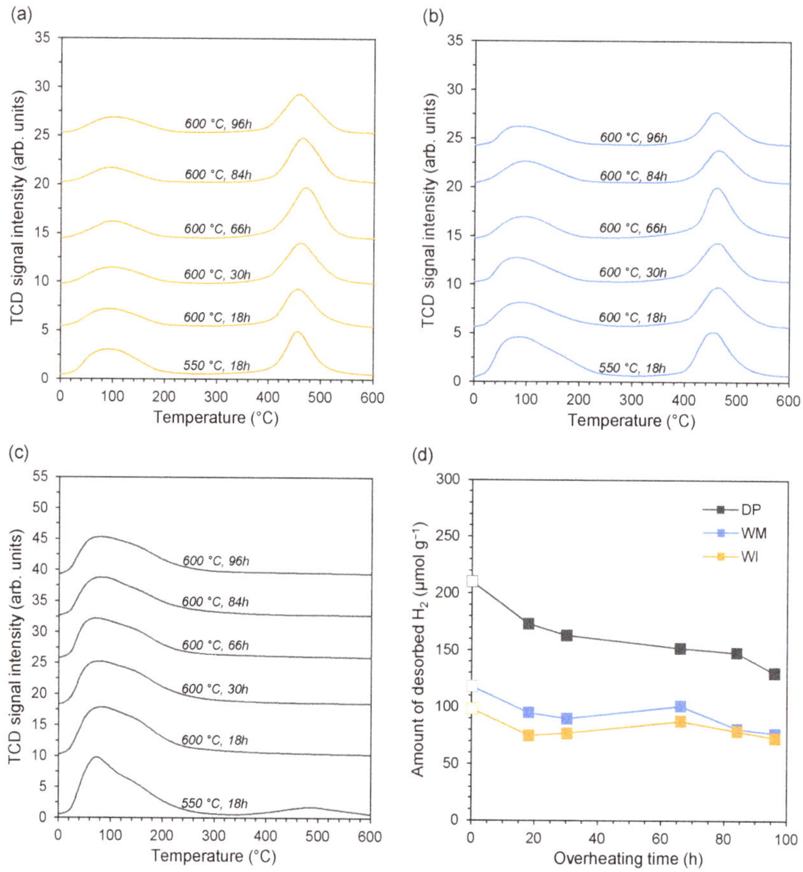

Figure 7. H_2-TPD profiles of the (**a**) WI catalyst, (**b**) WM catalyst, and (**c**) DP catalyst after reduction in hydrogen flow at 550 °C for 18 h and after overheating in hydrogen flow at 600 °C for 18, 30, 66, 84, and 96 h. (**d**) Changes in the amount of H_2 desorbed from the surface of the studied catalysts during long-term stability test.

The changes in the amount of desorbed H_2 from the surface of the tested catalysts during their long-term heat treatment are shown in Figure 7d. The observed trend of changes in the amount of desorbed H_2 is consistent with the observed slight changes in desorption profiles as a function of overheating time. For all the tested catalysts, the H_2 amount after overheating at 600 °C for 18 h decreased by ca. 20% compared to the amount desorbing after reduction at 550 °C for 18 h. Then, a slight decrease in the amount of desorbing hydrogen was observed during further long-term overheating at 600 °C, which finally, after 96 h, reached 3, 19, and 25% for WI, WM, and DP catalysts, respectively.

4. Discussion

It would seem that the most crucial feature of a catalyst is its activity. An equally, if not even more, critical feature for an industrial catalyst is its stability. The possibility of long-term operation of the catalyst in severe conditions of industrial processes without the need for regeneration or replacement is a basic requirement for their implementation. To fully assess the applicability of the catalyst, comprehensive characterization is required, including determination of its activity and investigation of its kinetic properties and stability of its operation under industrial conditions. The object of research presented in this paper was a cobalt catalyst promoted by cerium and barium for the ammonia synthesis process. As shown in our previous papers [30–32], this catalyst exhibits high activity in the NH_3 synthesis reaction, even several times higher than the activity of a commercial iron catalyst. However, the key point necessary to determine the possibility of its use in an industrial ammonia synthesis reactor is to determine its stability during long-term operation. The research described in this work aimed at determining the catalyst stability using the accelerated aging method, i.e., the only method that allows simulating in the laboratory the long-term catalyst operation in real reactor conditions. The obtained results helped us find the answer to the question about the factors determining changes in the activity of this catalyst during operation and indicate a potential mechanism for the deactivation of this catalyst.

Analyzing the results of the textural and structural characteristics of the catalyst, it can be concluded that the tested catalyst is highly resistant to high temperatures. The lack of significant changes in the size of the specific surface area (Table 1), morphology of the catalyst particles (Figure 3), distribution of elements on its surface (Figures 4–6), and chemisorption properties of the cobalt surface (Figure 7) during long-term heating at a temperature of 600 °C suggests that stable operation of the catalyst is possible in the ammonia synthesis reactor in the temperature range of 400–470 °C without the risk of losing its beneficial properties over time, which is confirmed by the data presented in Figure 1. However, the mechanism of catalyst deactivation by its slow sintering cannot be completely ruled out based on time-limited catalyst stability tests on a laboratory scale. Among the known mechanisms of catalyst deactivation, catalyst sintering seems to be the most likely cause of the loss of activity of a cobalt catalyst during its long-term operation, similar to a commercial fused iron catalyst. Despite its very high stability and exceptionally long lifespan (15 years or more), the Fe catalyst sinters slowly and requires periodic replacement. Important factors determining the lifespan of a cobalt catalyst are also resistance to poisoning, the possibility of regeneration in case of reversible poisoning (e.g., with oxygen compounds), and mechanical resistance. When comprehensively assessing the lifespan of a cobalt catalyst, it is necessary to consider poisoning tests and mechanical strength measurements for the formed catalyst.

The analysis of XRPD data draws attention to the presence of two crystalline forms of cobalt in the studied catalyst: hexagonal close-packed (hcp) and face-centered cubic (fcc). The proportion of these two phases changed when the catalyst was subjected to overheating at 600 °C, which was visible as a change in the intensity of reflections coming from these two phases (Figure 2). According to the literature [41–43], at a temperature of ca. 420 °C, a reversible hcp → fcc phase transformation occurs, i.e., at the temperature typical for the ammonia synthesis process. It is also known that the catalytic activity of cobalt in the ammonia synthesis reaction is strongly dependent on the form in which cobalt occurs in the catalyst [41]. The ability of N_2 and H_2 to adsorb on the cobalt surface depends on its phase. Hence, the NH_3 synthesis rates obtained for cobalt's hcp and fcc phases differ significantly. Rambeau's studies [41] showed that when conducting the process on cobalt with the hcp structure, the activity is twice as high as on cobalt with the fcc structure. Therefore, the hcp → fcc transformation is of particular importance in this case. The temperature of the ammonia synthesis process for a cobalt catalyst is close to the temperature of the hcp → fcc transformation of cobalt. Hence, this phenomenon may influence the efficiency of the NH_3 synthesis and the stability of catalyst operation at high temperatures over time.

The analysis of the obtained results also indicates that the method of introducing the Ba promoter does not impact the final properties, activity, and stability of the obtained CoCeBa catalyst. This is an extremely valuable observation from the point of view of the preparation of the tested catalyst. The introduction of the promoter is the last but crucial stage of catalyst synthesis. The presence of the Ba promoter in this catalyst is essential because it has an activating effect on cobalt and ensures good electron-donating properties of the catalyst surface required to conduct the NH_3 synthesis reaction effectively. The lack of a significant impact of the Ba addition method on the properties of the catalyst makes it possible to select the most appropriate method depending on the needs. Nowadays, much attention is paid to improving catalyst synthesis methods to achieve the highest possible activity of a given catalytic material in a specific process. Studies are focused on modifying synthesis methods, but more and more often turning into very complicated, sophisticated, and expensive methods that can be used to obtain small amounts of a highly active catalyst in a very time-consuming and/or energy-consuming process. However, the scale on which a given catalyst is to be finally used should always be considered. In the case of large-scale industrial processes, such as ammonia synthesis, a large amount of catalyst (several dozen tonnes) is required to be loaded into a single ammonia reactor. Hence, the catalyst production method should be relatively simple and easy to scale up. Considering the large-scale production of the CoCeBa catalyst for industrial purposes, the wet mixing method seems to be the most advantageous due to its simplicity and the lowest energy demand compared to other methods studied in this work (wet impregnation, deposition–precipitation).

5. Conclusions

Extended studies of the stability of the promoted cobalt catalyst for ammonia synthesis were carried out. The accelerated aging method was used to simulate the deterioration of catalyst performance during industrial operation. The catalyst was subjected to overheating for 436 h in a flow of a stoichiometric H_2/N_2 mixture at a temperature of 600 °C, i.e., a temperature much higher than the temperature of standard catalyst operation in the NH_3 synthesis reactor. The obtained results indicated that the CoCeBa catalyst is strongly resistant to overheating. The decline in catalyst activity during the long-term stability test was less than 10%. Textural and structural characteristics revealed the lack of significant changes in the specific surface area, morphology of the catalyst particles, and surface distribution of elements over time. Only a slight decrease in the amount of desorbed H_2 from the cobalt surface during their long-term heat treatment was observed. However, the nature of the adsorption sites on the catalyst surface did not change. The stable operation of the catalyst is possible in the ammonia synthesis reactor in the temperature range of 400–470 °C without the risk of losing its beneficial catalytic properties over time. The lack of a significant impact of the Ba addition method on the properties of the catalyst makes it possible to select the most appropriate method from the point of view of simplicity, costs, and energy consumption in the catalyst production process.

Supplementary Materials: The following supporting information can be downloaded at: https://www.mdpi.com/article/10.3390/en16237787/s1, Table S1: Physicochemical properties of the studied catalyst precursors; Table S2: Comparison of the catalytic performance for NH3 synthesis of different catalysts; Table S3: Unit cell parameters and crystallite size of BaCeO3 phase in the CoCeBa catalyst precursors; Table S4: Unit cell parameters and crystallite size of BaCeO3 phase in the CoCeBa catalysts after overheating in H2 (600 °C, 96 h). Figure S1: N_2 adsorption–desorption isotherms for (a) WI, (b) DP, (c) WM catalyst precursors obtained based on BJH adsorption isotherm model; Figure S2: Pore size distribution curves for (a) WI, (b) DP, (c) WM catalyst precursors obtained based on BJH adsorption isotherm model; Figure S3: XRPD patterns of the WI, DP, WM catalyst precursors; Figure S4: Arrhenius plots for the rate of ammonia synthesis at 6.3 MPa over the WI, DP, WM catalysts.

Author Contributions: Conceptualization, M.Z., H.R. and W.R.-P.; methodology, M.Z., H.R. and W.R.-P.; investigation, M.Z., H.R., W.P., W.R., A.O., K.S. and W.R.-P.; writing—original draft preparation, M.Z.; writing—review and editing, M.Z., H.R. and W.R.-P.; visualization, H.R. and K.S.; supervision, M.Z. All authors have read and agreed to the published version of the manuscript.

Funding: This research received no external funding.

Data Availability Statement: Data are contained within the article and supplementary materials.

Conflicts of Interest: The authors declare no conflict of interest.

References

1. Vezirogl'u, T.N.; Barbir, F. Hydrogen: The wonder fuel. *Int. J. Hydrogen Energy* **1992**, *17*, 391–404. [CrossRef]
2. Amin, M.; Shah, H.H.; Fareed, A.G.; Khan, W.U.; Chung, E.; Zia, A.; Farooqi, Z.U.; Lee, C. Hydrogen production through renewable and non-renewable energy processes and their impact on climate change. *Int. J. Hydrogen Energy* **2022**, *47*, 33112–33134. [CrossRef]
3. Singh, S.; Jain, S.; Ps, V.; Tiwari, A.K.; Nouni, M.R.; Pandey, J.K.; Goel, S. Hydrogen: A sustainable fuel for future of the transport sector. *Renew. Sustain. Energy Rev.* **2015**, *51*, 623–633. [CrossRef]
4. Nikolaidis, P.; Poullikkas, A. A comparative overview of hydrogen production processes. *Renew. Sustain. Energy Rev.* **2017**, *67*, 597–611. [CrossRef]
5. Jiang, L.-W.; Huang, Y.; Zou, Y.; Meng, C.; Xiao, Y.; Liu, H.; Wang, J.-J. Boosting the Stability of Oxygen Vacancies in α-$Co(OH)_2$ Nanosheets with Coordination Polyhedrons as Rivets for High-Performance Alkaline Hydrogen Evolution Electrocatalyst. *Adv. Energy Mater.* **2022**, *12*, 2202351. [CrossRef]
6. Wu, Y.Z.; Huang, Y.; Jiang, L.-W.; Meng, C.; Yin, Z.-H.; Liu, H.; Wang, J.-J. Modulating the electronic structure of CoS_2 by Sn doping boosting urea oxidation for efficient alkaline hydrogen production. *J. Colloid Interface Sci.* **2023**, *642*, 574–583. [CrossRef]
7. Huang, Y.; Jiang, L.-W.; Liu, H.; Wang, J.-J. Electronic structure regulation and polysulfide bonding of Co-doped $(Ni, Fe)_{1+x}S$ enable highly efficient and stable electrocatalytic overall water splitting. *Chem. Eng. J.* **2022**, *441*, 136121. [CrossRef]
8. Acar, C.; Dincer, I. Review and evaluation of hydrogen production options for better environment. *J. Clean. Prod.* **2019**, *218*, 835–849. [CrossRef]
9. Lucentini, I.; Garcia, X.; Vendrell, X.; Llorca, J. Review of the Decomposition of Ammonia to Generate Hydrogen. *Ind. Eng. Chem. Res.* **2021**, *60*, 18560–18611. [CrossRef]
10. Lan, R.; Irvine, J.T.S.; Tao, S. Ammonia and related chemicals as potential indirect hydrogen storage materials. *Int. J. Hydrogen Energy* **2012**, *37*, 1482–1494. [CrossRef]
11. Klerke, A.; Christensen, C.H.; Nørskov, J.K.; Vegge, T. Ammonia for hydrogen storage: Challenges and opportunities. *J. Mater. Chem.* **2008**, *18*, 2304–2310. [CrossRef]
12. Berwal, P.; Kumar, S.; Khandelwal, B. A comprehensive review on synthesis, chemical kinetics, and practical application of ammonia as future fuel for combustion. *J. Energy Inst.* **2021**, *99*, 273–298. [CrossRef]
13. Herbinet, O.; Bartocci, P.; Grinberg Dana, A. On the use of ammonia as a fuel—A perspective. *Fuel Commun.* **2022**, *11*, 100064. [CrossRef]
14. Elbaz, A.M.; Wang, S.; Guiberti, T.F.; Roberts, W.L. Review on the recent advances on ammonia combustion from the fundamentals to the applications. *Fuel Commun.* **2022**, *10*, 100053. [CrossRef]
15. Valera-Medina, A.; Amer-Hatem, F.; Azad, A.K.; Dedoussi, I.C.; de Joannon, M.; Fernandes, R.X.; Glarborg, P.; Hashemi, H.; He, X.; Mashruk, S.; et al. Review on Ammonia as a Potential Fuel: From Synthesis to Economics. *Energy Fuels* **2021**, *35*, 6964–7029. [CrossRef]
16. Zhang, L.; Liang, J.; Wang, Y.; Mou, T.; Lin, Y.; Yue, L.; Li, T.; Liu, Q.; Luo, Y.; Li, N.; et al. High-Performance Electrochemical NO Reduction into NH_3 by MoS_2 Nanosheet. *Angew. Chem. Int. Ed.* **2021**, *60*, 25263. [CrossRef] [PubMed]
17. Liu, H. *Ammonia Synthesis Catalyst: Innovations and Practice*; World Scientific Publishing Co. Pte. Ltd.: Singapore; Chemical Industry Press: Beijing, China, 2013; pp. 686–705.
18. Hughes, R. *Deactivation of Catalysts*; Academic Press: London, UK, 1984.
19. Moulijn, J.A.; van Diepen, A.E.; Kapteijn, F. Catalyst deactivation: Is it predictable? What to do? *Appl. Catal. A Gen.* **2001**, *212*, 3–16. [CrossRef]
20. Argyle, M.D.; Bartholomew, C.H. Heterogeneous Catalyst Deactivation and Regeneration: A Review. *Catalysts* **2015**, *5*, 145–269. [CrossRef]
21. Bartholomew, C.H. Mechanisms of catalyst deactivation. *Appl. Catal. A Gen.* **2001**, *212*, 17–60. [CrossRef]
22. Bartholomew, C.H.; Farrauto, R.J. *Fundamentals of Industrial Catalytic Processes*, 2nd ed.; John Wiley & Sons, Inc.: Hoboken, NJ, USA, 2006.
23. Birtill, J.J. But will it last until the shutdown? Deciphering catalyst decay! *Catal. Today* **2003**, *81*, 531–545. [CrossRef]
24. Scott, S.L. A Matter of Life(time) and Death. *ACS Catal.* **2018**, *8*, 8597–8599. [CrossRef]
25. Pernicone, N. Methods for laboratory-scale evaluation of catalyst life in industrial plants. *Appl. Catal.* **1985**, *15*, 17–31. [CrossRef]

26. Liu, H.-Z.; Li, X.-N.; Hu, Z.-N. Development of novel low temperature and low pressure ammonia synthesis catalyst. *Appl. Catal. A* **1996**, *142*, 209–222. [CrossRef]
27. Pernicone, N.; Ferrero, F.; Rossetti, I.; Forni, L.; Canton, P.; Riello, P.; Fagherazzi, G.; Signoretto, M.; Pinna, F. Wustite as a new precursor of industrial ammonia synthesis catalysts. *Appl. Catal. A* **2003**, *251*, 121–129. [CrossRef]
28. Lin, B.; Guo, Y.; Lin, J.; Ni, J.; Lin, J.; Jiang, L.; Wang, Y. Deactivation study of carbon-supported ruthenium catalyst with potassium promoter. *Appl. Catal. A* **2017**, *541*, 1–7. [CrossRef]
29. Kowalczyk, Z.; Jodzis, S.; Raróg, W.; Zieliński, J.; Pielaszek, J. Effect of potassium and barium on the stability of a carbon-supported ruthenium catalyst for the synthesis of ammonia. *Appl. Catal. A* **1998**, *173*, 153–160. [CrossRef]
30. Raróg-Pilecka, W.; Karolewska, M.; Truszkiewicz, E.; Iwanek, E.; Mierzwa, B. Cobalt catalyst doped with cerium and barium obtained by Co-precipitation method for ammonia synthesis process. *Catal. Lett.* **2011**, *141*, 678–684. [CrossRef]
31. Karolewska, M.; Truszkiewicz, E.; Mierzwa, B.; Kępiński, L.; Raróg-Pilecka, W. Ammonia synthesis over cobalt catalysts doped with cerium and barium. Effect of the ceria loading. *Appl. Catal. A* **2012**, *445–446*, 280–286. [CrossRef]
32. Tarka, A.; Patkowski, W.; Zybert, M.; Ronduda, H.; Wieciński, P.; Adamski, P.; Sarnecki, A.; Moszyński, D.; Raróg-Pilecka, W. Synergistic Interaction of Cerium and Barium-New Insight into the Promotion Effect in Cobalt Systems for Ammonia Synthesis. *Catalysts* **2020**, *10*, 658. [CrossRef]
33. Kowalczyk, Z. Effect of potassium on the high-pressure kinetics of ammonia synthesis over fused iron catalysts. *Catal. Lett.* **1996**, *37*, 173–179. [CrossRef]
34. Lin, B.; Liu, Y.; Heng, L.; Wang, X.; Ni, J.; Lin, J.; Jiang, L. Morphology Effect of Ceria on the Catalytic Performances of Ru/CeO_2 Catalysts for Ammonia Synthesis. *Ind. Eng. Chem. Res.* **2018**, *57*, 9127–9135. [CrossRef]
35. Han, W.; Li, Z.; Liu, H. $La_2Ce_2O_7$ supported ruthenium as a robust catalyst for ammonia synthesis. *J. Rare Earths* **2019**, *37*, 492–499. [CrossRef]
36. Jafari, A.; Ebadi, A.; Sahebdelfar, S. Effect of iron oxide precursor on the properties and ammonia synthesis activity of fused iron catalysts. *Reac. Kinet. Mech. Cat.* **2019**, *126*, 307–325. [CrossRef]
37. Zhou, Y.; Ma, Y.; Lan, G.; Tang, H.; Han, W.; Liu, H.; Li, Y. A highly stable and active mesoporous ruthenium catalyst for ammonia synthesis prepared by a $RuCl_3$/SiO_2-templated approach. *Chin. J. Catal.* **2019**, *40*, 114–123. [CrossRef]
38. Zybert, M.; Karasińska, M.; Truszkiewicz, E.; Mierzwa, B.; Raróg-Pilecka, W. Properties and activity of the cobalt catalysts for NH_3 synthesis obtained by co-precipitation – the effect of lanthanum addition. *Pol. J. Chem. Technol.* **2015**, *17*, 138–143. [CrossRef]
39. Lin, B.; Liu, Y.; Heng, L.; Ni, J.; Lin, J.; Jiang, L. Effect of barium and potassium promoter on Co/CeO_2 catalysts in ammonia synthesis. *J. Rare Earths* **2018**, *36*, 703–707. [CrossRef]
40. Ronduda, H.; Zybert, M.; Patkowski, W.; Sobczak, K.; Moszyński, D.; Albrecht, A.; Sarnecki, A.; Raróg-Pilecka, W. On the effect of metal loading on the performance of Co catalysts supported on mixed MgO–La_2O_3 oxides for ammonia synthesis. *RSC Adv.* **2022**, *12*, 33876–33888. [CrossRef]
41. Rambeau, G.; Jorti, A.; Amariglio, H. Catalytic activity of a cobalt powder in NH_3 synthesis in relation with the allotropic transformation of the metal. *J. Catal.* **1985**, *94*, 155–165. [CrossRef]
42. Ray, A.E.; Smith, S.R.; Scofield, J.D. Study of the phase transformation of cobalt. *Phase Equilibr.* **1991**, *12*, 644–647. [CrossRef]
43. Matsumoto, H. Shift of $\alpha \rightarrow \beta$ transformation temperature of cobalt with thermal cycling. *J. Mater. Sci. Lett.* **1993**, *12*, 969–970. [CrossRef]
44. Ronduda, H.; Zybert, M.; Patkowski, W.; Moszyński, D.; Albrecht, A.; Sobczak, K.; Małolepszy, A.; Raróg-Pilecka, W. Co nanoparticles supported on mixed magnesium–lanthanum oxides: Effect of calcium and barium addition on ammonia synthesis catalyst performance. *RSC Adv.* **2023**, *13*, 4787–4802. [CrossRef] [PubMed]
45. Schlögl, R. Catalytic Synthesis of Ammonia—A "Never-Ending Story"? *Angew. Chem. Int. Ed.* **2003**, *42*, 2004–2008. [CrossRef] [PubMed]
46. Humphreys, J.; Lan, R.; Tao, S. Development and Recent Progress on Ammonia Synthesis Catalysts for Haber–Bosch Process. *Adv. Energy Sustain. Res.* **2021**, *2*, 2000043. [CrossRef]
47. Ertl, G. Reactions at Surfaces: From Atoms to Complexity (Nobel Lecture). *Angew. Chem. Int. Ed.* **2008**, *47*, 3524–3535. [CrossRef] [PubMed]
48. Cao, A.; Bukas, V.J.; Shadravan, V.; Wang, Z.; Li, H.; Kibsgaard, J.; Chorkendorff, I.; Nørskov, J.K. A spin promotion effect in catalytic ammonia synthesis. *Nat. Commun.* **2022**, *13*, 2382. [CrossRef]
49. Ertl, G. Elementary steps in ammonia synthesis: The surface approach. In *Catalytic Ammonia Synthesis. Fundamentals and Practice*, 1st ed.; Jennings, J.R., Ed.; Springer Science+Business Media: New York, NY, USA, 1991; pp. 109–132.
50. Ronduda, H.; Zybert, M.; Patkowski, W.; Tarka, A.; Jodłowski, P.; Kępiński, L.; Sarnecki, A.; Moszyński, D.; Raróg-Pilecka, W. Tuning the catalytic performance of Co/Mg-La system for ammonia synthesis via the active phase precursor introduction method. *Appl. Catal. A* **2020**, *598*, 117553. [CrossRef]
51. Ronduda, H.; Zybert, M.; Dziewulska, A.; Patkowski, W.; Sobczak, K.; Ostrowski, A.; Raróg-Pilecka, W. Ammonia synthesis using Co catalysts supported on MgO–Nd_2O_3 mixed oxide systems: Effect of support composition. *Surf. Interfaces* **2023**, *36*, 102530. [CrossRef]

energies

Article

Designing Sustainable Ethanol Oxidation Catalysts: The Role of Graphene Oxide in NiCuGO Composite Material

Marta Wala-Kapica, Magdalena Szewczyk and Wojciech Simka *

Department of Inorganic Chemistry, Analytical Chemistry and Electrochemistry, Faculty of Chemistry, Silesian University of Technology, 44-100 Gliwice, Poland; marta.wala-kapica@polsl.pl (M.W.-K.); magdsze594@student.polsl.pl (M.S.)
* Correspondence: wojciech.simka@polsl.pl

Abstract: The growing world population with the growth of civilization is causing the demand for electric energy to increase every year. For this reason, new energy sources such as fuel cells are becoming more and more needed, especially when they can use renewable fuel such as ethanol. This simple organic alcohol can be easily produced in a fermentation process using organic waste. Its oxidation might be used as a source for electricity; however, due to the lack of proper electrocatalytic materials, such a solution is not popular. A simple method of NiCuGO composite preparation via electrodeposition from a water-based solution containing graphene oxide suspension is proposed. The activity of the prepared material is proven, with higher current densities observed for the composite powder. The highest peak current density is observed for NiCuGO deposited with a higher current density. The observed i_{pA} of 8.6 mA cm^{-2} has been higher than that reported by other researchers.

Keywords: ethanol oxidation; fuel cell; nickel; copper; graphene oxide; composite; electrodeposition

1. Introduction

Fuel cells stand as a leading contender in the quest for green energy solutions, offering high efficiency and a low environmental impact. However, the widespread adoption of fuel cells is hindered by the limited performance and high cost of materials employed in their construction, particularly within the anodic components [1–3]. The continuous pursuit of alternative materials that can enhance the efficiency, stability, and cost-effectiveness of fuel cells has led to the exploration of various composite materials [1,3].

Simple alcohols have increasingly received attention as potential fuel sources due to their practicality as energy providers. Their attraction primarily lies in their high energy densities, low toxicity, and ease of both storage and transportation. Among these alcohols, ethanol stands out prominently due to its remarkable suitability arising from the straightforward synthesis via the biological process of fermentation. This unique characteristic endows ethanol with a renewable nature. Notably, virtually any organic source, especially one abundant in carbohydrates, can be utilized as feedstock for ethanol production, reinforcing its renewable quality. This inherent characteristic underscores the potential for an eco-friendly and sustainable energy source, fostering continued interest and exploration in alcohols as a viable fuel alternative [3–7].

Among the multitude of materials under investigation, nickel and copper emerge as promising candidates due to their catalytic properties and abundance [8–14]. Additionally, the integration of graphene oxide—a remarkable two-dimensional carbon-based material known for its exceptional conductivity, and large surface area—presents an exciting avenue for enhancing the performance of fuel cell components [13,15–20].

Nickel-based materials have been proven to be active towards ethanol oxidation. The synergic effect of its mixing with copper has also been reported to be beneficial for hydrogen evolution [21,22], oxygen evolution [23], oxygen reduction [24], methanol oxidation [25], urea oxidation, [26,27] ethanol oxidation [28] and dehydrogenation [29,30]. The amalgamation of these materials aims to leverage the synergistic effects of their individual properties, capitalizing on the catalytic activity of nickel and copper while harnessing the conductivity and structural reinforcement provided by graphene oxide. The addition of copper to Ni-based catalysts has been proven to increase its activity due to the electronic effect, which is related to the vacancy in the filling of the *d* band of nickel and in the suppression of the formation of less active γ-NiOOH [31,32]. Another explanation of the higher activity of Ni catalysts doped with Cu is a bifunctional mechanism, where the Cu atoms are active centers for OH^- ion adsorption [29,33] and increased immunity toward CO poisoning caused by the presence of Cu [27]. Such activity could be even further increased via the preparation of composite material, consisting of both Ni and Cu metals with the addition of carbon nanomaterial, which would not only improve the conductivity of the final material, but also increase its active surface area and the particles' dispersion during electrodeposition [15,34–41].

This article presents a comprehensive study detailing the synthesis and characterization of a new composite material consisting of nickel, copper, and graphene oxide for ethanol fuel cell anodes. The NiCuGO composite prepared via the electrodeposition of a water suspension, which has been shown to be active towards methanol [42] and urea [43], is now examined for ethanol oxidation. The addition of graphene oxide has been proven to increase the surface area of the proposed material and increase its activity toward methanol and urea oxidation. Since ethanol is a much more complicated particle and the breaking of the strong intercarbon bond is not easy, this article focusses on an examination of the activity of the proposed material towards its oxidation and the role of graphene oxide in the process.

The proposed material was synthesized via electrodeposition from a water-based electrolytic bath, offering a streamlined, rapid, and environmentally friendly process. This methodology stands out for its exclusion of toxic organic substances, such as hydrazine or NH_4F, commonly employed in the synthesis of similar materials [35,44–47]. Moreover, the electrodeposition process provides a versatile means to tailor the final material by adjusting various process parameters, including current density and electrolyte composition [41,48–51]. This adaptability allows for precise control over the material's characteristics, demonstrating a flexible and efficient approach to material design and synthesis.

The best results were observed for a NiCuGO composite electrodeposited with a current density of 20.8 A dm^{-2}. The peak current density for ethanol oxidation was 8.6 mA cm^{-2} at a peak potential of 0.66 V vs. Ag | $AgCl_{(sat.KCl)}$ and 9.1 mA cm^{-2} at a peak potential of 0.7 V vs. Ag | $AgCl_{(sat.KCl)}$.

2. Materials and Methods

All used chemicals were of analytical purity, and were manufactured by Avantor (Gliwice, Poland).

The proposed method of composite powder preparation has been described elsewhere [42]. Briefly, to a water-based solution containing nickel and copper salts with the addition of citric acid as a complexing agent suspension, 0.5 g L^{-1} of graphene oxide (Graphene Supermarket, Ronkonkoma, NY, USA) was added. Using the power source with a current density of 15.6 A dm^{-2} and 20.8 A dm^{-2} working in constant-current mode for 10 and 5 min, respectively, the proposed materials were electrodeposited in the forms of powders, which were taken off the base metal using ultrasounds, dried and cleaned. Depending on the used current density and the electrolyte ingredients, samples were named, as shown in Table 1.

Table 1. Clarification of samples codes used further in the text.

Sample Name	GO Addition, g L^{-1}	i, A dm^{-2}
NiCu15	-	15.6
NiCu20		20.8
NiCuGO15	0.5	15.6
NiCuGO20		20.8

The electrochemical activity of prepared materials was assessed using the AUTOLAB PGSTAT302N potentiostat with NOVA 2.1 software (Methrom, Herisau, Switzerland). For electrochemical measurements, a three-electrode system was used, with glassy carbon as a counter electrode and $Ag|AgCl_{(sat.KCl)}$ as a reference electrode, which is why all the potentials mentioned in this paper will be given in the reference to this electrode.

All mentioned measurements were made on stationary working electrodes, which were prepared via the deposition of 5 μL of catalytic ink on a polished glassy carbon electrode (ϕ 5 mm). Catalytic ink was prepared by mixing the prepared powder (1 mg of powder per 30 μL of ink) with Nafion® and Isopropanol solutions (1:10 v/v), followed by ultrasound mixing until deep black catalytic ink was prepared.

The activity of the proposed material was examined using cyclic voltammetry in the voltage range of 0–0.7 V vs. $Ag|AgCl_{(sat.KCl)}$, linear weep voltammetry and chronoamperometry.

The best activity of the proposed powders was observed in 1 M KOH, which was used as a supporting electrolyte for ethanol. The concentration of ethanol was experimentally chosen via a series of CV experiments in solutions with various (1–5 M) ethanol concentrations.

Scanning electron microscopy (SEM) was conducted using Hitachi S-4000N (Hitachi, Tokyo, Japan) with an exceeding voltage of 15 kV. Based on SEM images, particle size analysis was conducted using the ImageJ 1.5 (NIH, Bethesda, MD, USA) graphical software. The size of the particle edges was measured at 50 points for each modification.

3. Results and Discussion

The proposed powders were examined using XRD [42], XPS and Raman methods [43]. They were proven to be $Cu_{0.21}Ni_{0.79}$ alloys with the presence of GO. On the surface, powders were covered with metal oxides or salts.

The morphology of obtained powders was typical for metallic powders electrodeposited with a high current density. Dendrite- or coral-like particles were formed due to the electrodeposition process taking place at the same time with hydrogen evolution. Gas bubbles forming on the surface of the substrate played the role of a dynamic template, which leading to the highly developed surface with a highly electroactive surface [52–55]. Such a quality is very beneficial for the potential use of electrodeposits as catalysts, since this process has a strong surface characteristic, and its yield is strictly related to the size of the electroactive surface. The dendritic morphology of catalyst is related to the strengthened electric field distribution, which prompts local OH^- enrichment in the electrical double layer (EDL), so the catalytic effect of the material is reinforced [56].

The morphology of the prepared working electrode is shown in Figure 1.

Based on the SEM images of proposed samples, the distribution of particles size was evaluated (Figure 2). From these plots, we can conclude that the usage of a higher current density leads to the electrodeposition of smaller particles, which suggests that the speed of nucleus formation is higher than the speed of nucleus propagation.

Interestingly, the addition of graphene oxide into the electrolyte led to the formation of smaller particles, despite the use of a smaller current density, which suggests that the presence of GO in the electrolytic bath changed the deposition mechanism more strongly than just its incorporation into the final material. The smallest particles were observed for NiCuGO20 powder, which suggests that this powder should have the highest electroactive active surface (ECSA), which has been already proven [42]. Its superior activity has already been proven for 1 M KOH solution with and without the addition of methanol and urea.

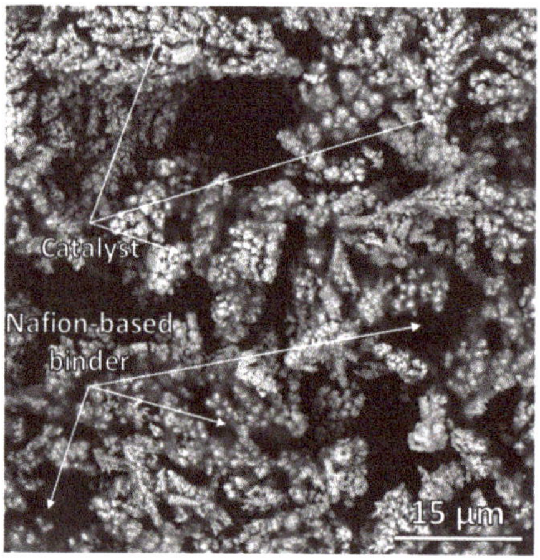

Figure 1. SEM image of NiCuGO20-based working electrode.

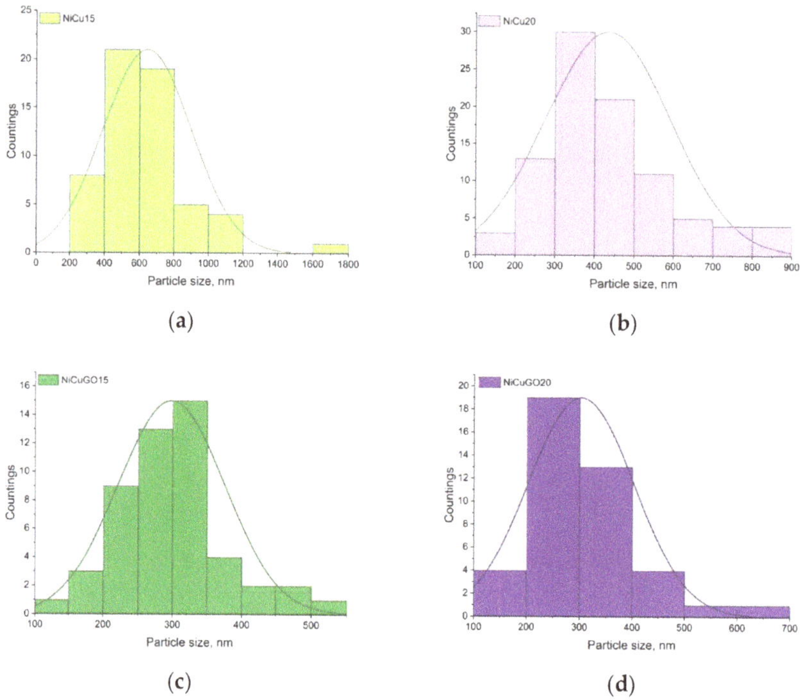

Figure 2. Particle size distribution of prepared materials: (**a**) NiCu15, (**b**) NiCu20, (**c**) NiCuGO15 and (**d**) NiCuGO20.

3.1. *Influence of Ethanol Concentrations*

Prior to investigating the electrochemical activity of the proposed materials for ethanol oxidation, we conducted preliminary assessments of their performance in various KOH solutions to determine the optimal electrolyte concentration. Figure 3 illustrates the representative current response recorded for NiCuGO20. As the KOH concentration increased from 0.5 to 1 M, there was a noticeable rise in the peak current density. However, it is noteworthy that a further increase in KOH concentration from 1 to 2 M did not result in a corresponding increase in current density. On the contrary, the observed peak was smaller than that recorded in the 1 M solution. Since this trend was consistent across all prepared powders, we selected a 1 M KOH concentration as the supporting electrolyte for subsequent studies.

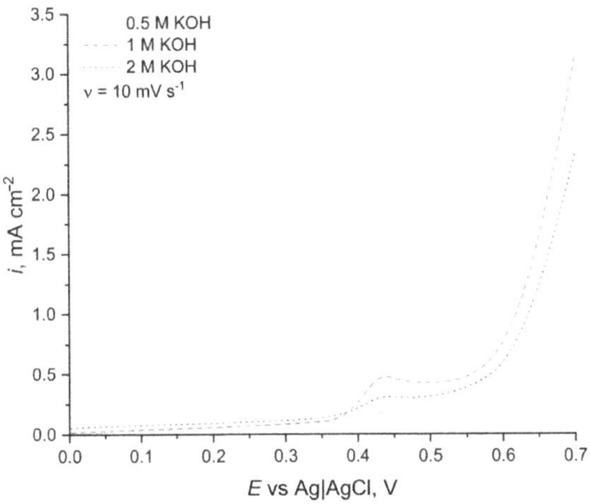

Figure 3. Linear sweep voltammetry results for NiCuGO20 powder.

3.2. *Activity towards Ethanol Oxidation*

The assessment of proposed material activity started with an investigation of the influence of ethanol concentration on the activity of the proposed materials.

The ethanol concentration for further electroactivity studies was chosen based on a series of voltametric experiments. Briefly, 1 M KOH solutions with the addition of 1, 2 and 5 M of ethanol were prepared, and the activity of the proposed samples were examined, as presented in Figure 4.

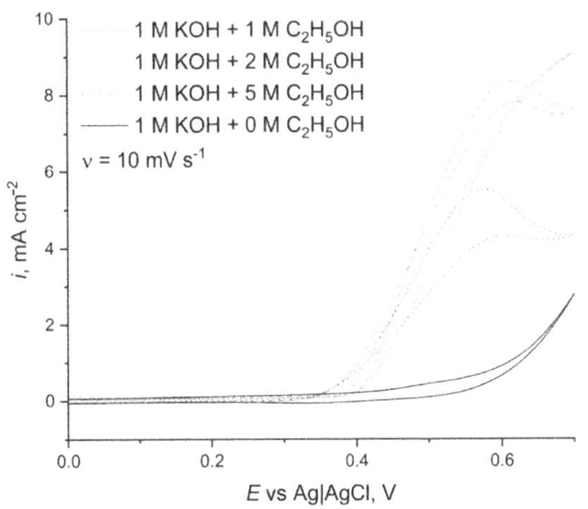

Figure 4. Cyclic voltammograms of NiCuGO20 powder in 1 M KOH with the addition of various ethanol concentrations.

The highest current density was observed for 1 M ethanol solution. A lower concentration of this alcohol resulted in lower peak current densities, and interestingly, a further increase also led to lower peak current densities, which was probably related to the usage of all active centers by 1 M of this molecule.

3.3. *Activity towards Ethanol Oxidation*

All the proposed powders showed activity towards ethanol oxidation, as shown in Figure 5. During the forwards scan, an oxidation peak was visible, which was related to the reactions [13,57–59]:

$$NiOOH + C_2H_5OH \rightarrow Ni(OH)_2 + products \quad (1)$$

$$CuOOH + C_2H_5OH \rightarrow Cu(OH)_2 + products \quad (2)$$

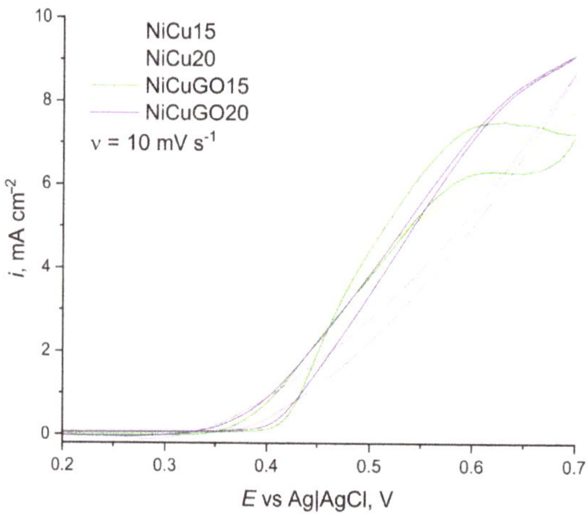

Figure 5. Cyclic voltammogram of prepared powders; 1 M KOH + 1 M ethanol.

Unfortunately, we cannot conclude what the products of such a reaction are. Ethanol oxidation can take place via two different mechanisms: C1, leading to CO_2 as a reaction product, or the C2 pathway, leading mostly to acetic acid and acetaldehyde [60,61]. We have not examined volatile or liquid products in the reaction mixture, which is why no further conclusions can be considered and why further studies of the GO influence on the materials' activity are needed.

Powder NiCu20 has shown slightly different behavior compared to that of the other examined materials, since in its voltammograms no clear oxidation peak was visible. Such a difference might be related to the different mechanisms of ethanol oxidation and thus the lack of a stable diffusion layer causing the characteristic CV shape for oxidation reactions.

Nevertheless, we can conclude that its activity towards ethanol oxidation is lower than its composite equivalent, since its current outcomes were lower in the whole voltage range in which the CV scans were conducted.

The highest activity, proven by the highest peak current density, with as value of 8.6 mA cm^{-2}, was been observed for NiCuGO20 composite powder. Interestingly, the peak current density observed for NiCuGO15 composite is lower than the one observed for its alloy equivalent (see Table 2). Such a difference shows once again that the mechanism of the influence of graphene oxide on material activity is not so easy to determine, and that other parameters, such as the mean particle size, related to the deposition current density, can change the final material's activity even more strongly than can just the presence of graphene oxide.

Table 2. Comparison of observed current outcomes at oxidation peaks and at reverse potential (0.7 V vs. Ag | $AgCl_{(sat.KCl)}$).

Sample	E_{pA}, V	i_{pA}, mA cm^{-2}	$i_{@Erev}$, mA cm^{-2}
NiCu15	0.65	7.05	7.75
NiCu20	-	-	8.70
NiCuGO15	0.63	7.5	7.25
NiCuGO20	0.66	8.6	9.10

E_{pA}—anodic peak potential; i_{pA}—anodic peak current density; $i_{@Erev}$—peak current density at the reverse potential.

Similar materials described in the literature have shown lower peak current densities, from 4.03 mA cm^{-2} observed for nanomaterial Ni20@Pd60Rh20 [62] through to 5.6 mA cm^{-2} for Ni-doped platinum [63]. As it is presented in Table 3 below, similar activity has been reported for electrodeposited NiCu material, with an *ipA* of 12.4 mA cm^{-2} [64]. Ni–Co composite deposited on reduced, nitrogen-doped GO has shown even higher activity, this being 64.2 mA cm^{-2} [65]. For NiCu nanorods, the recorded peak current density was 86.1 mA cm^{-2} [58]. A catalyst based on Ni, Cu and Fe_2O_3 showed even higher activity with an *ipA* of 101 mA cm^{-2} [66], similarly to mixed a Ni–NiO catalyst, which had a peak current density of 120 mA cm^{-2} [67]. Catalysts consisting of Pd doped with Cu [68] or Cd with the addition of reduced GO [69] have shown an *ipA* of 166 and 179 mA cm^{-2}, respectively. Other dopants used for Ni-based catalysts, such as Co or Cr_2O_3, have also led to an increase in the recorded peak current density up to 181 [57] and 327 mA cm^{-2} [70], respectively.

Table 3. Comparison of the proposed materials with the materials described in the literature.

Material	i_{pA}, mA cm^{-2}	Reference
NiCuGO20	8.6	This article
NiCo	180	[57]
NiCu NR	86.1	[58]
Ni_{20}@$Pd_{60}Rh_{20}$/C	4.0	[62]
PtNi/C	5.6	[63]
NiCu	12.4	[64]
NiCo@ErN-GO	64.2	[65]
$NiCuFe_2O_3$	101	[66]
Ni-NiO	120	[67]
PdCu	166	[68]
PdCdrGO	179	[69]
Ni-Cr_2O_3	327	[70]

NR—nanorods.

The proposed composite showed higher activity than that of nickel-based nanomaterials. However, it showed lower activity than that of a similar NiCu-based material, which was probably caused by the bigger size of the prepared particles, leading to a lower active surface. Despite the lower activity, the proposed catalyst can still be a good candidate as an ethanol oxidation catalyst due to its rapid synthesis and stability.

3.4. Stability during Ethanol Oxidation

As presented in Figure 6, all proposed materials were proven to be stable during 1800 s polarization to their peak potential. After the start of polarization, initially, the current density grew, contrastingly to the usually recorded drop related to the charging of the dielectric layer. In this case, the current response grew, which has also been reported for similar materials [71].

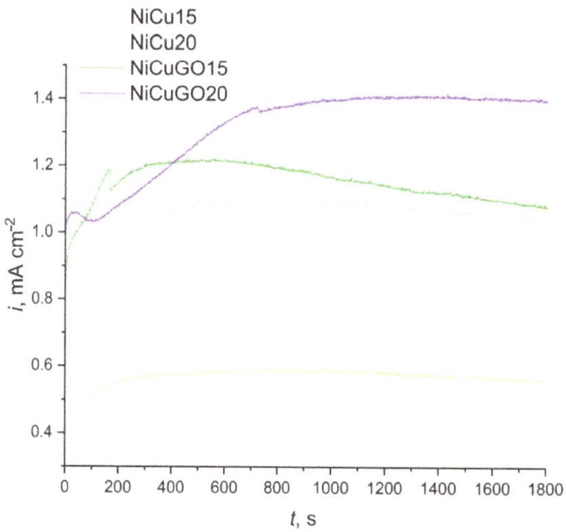

Figure 6. Results of chronoamperometric studies recorded for 1800 s at the anodic peak potential.

Such behavior is related to the fact that ethanol as an organic molecule consisting of two carbon atoms is not easy to be fully oxidized. The quality that makes it such a stable compound, making it a perfect fuel to store and transport, also makes it hard to fully oxidize. The stable bond between the two carbon atoms is very hard to break, which is why ethanol oxidation might need more time and would explain why after the beginning of polarization to the peak potential current the response grows with time until it stabilizes. The peak visible on the CA curves is probably related to the moment when the multistep ethanol oxidation reaction would have occurred with the highest efficiency. The lowering of current densities recorded at the peak potential might be related to the slight poisoning of the catalyst from the reaction products or to the mechanical detachment of the catalyst caused by the gaseous products' evolution. Once again, the highest activity was observed for NiCuGO20 composite, which proves that this modification is not only the most active during voltametric studies, but also that it preserves its activity during long-term examination.

The ethanol oxidation mechanism on NiCuGO composite catalysts is a complex process that involves multiple steps, with a key focus on the regeneration of active sites [72,73]. Notably, other researchers have emphasized that the activity of the system can be significantly enhanced through the incorporation of certain agents that improve electron conductivity [66,74,75]. The NiCuGO composite catalyst appears to benefit from the addition of graphene oxide, which plays a crucial role in facilitating electron transfer during the ethanol oxidation process. The high electron conductivity of GO contributes to the overall efficiency of the catalyst, making it a promising material for ethanol oxidation applications.

4. Conclusions

In the presented paper, a quick and easy method of electrocatalyst composite preparation is described. The properties of the final material can be tailored by changing the conditions of the electrodeposition process, which is a huge advantage of the proposed method. It has been proven that the addition of graphene oxide into the electrolytic bath strongly influences the properties of the final material, leading to its smaller particle sizes and higher activity. The best results in terms of ethanol oxidation were observed for NiCuGO20 composite, which shows that there is a strong correlation between the process parameters and the properties of the prepared material. Higher deposition current densities and the presence of GO in the electrolytic bath lead to the formation of more active particles.

Author Contributions: Conceptualization, M.W.-K. and W.S.; methodology, M.W.-K. and W.S.; software, M.W.-K.; formal analysis, W.S.; investigation, M.W.-K. and M.S.; resources, M.W.-K. and M.S.; data curation, M.W.-K. and M.S.; writing—original draft preparation, M.W.-K.; writing—review and editing, M.W.-K. and W.S.; supervision, W.S. All authors have read and agreed to the published version of the manuscript.

Funding: This research received no external funding.

Data Availability Statement: Data are contained within the article.

Conflicts of Interest: The authors declare no conflicts of interest.

References

1. Sequeira, C.A.C.; Cardoso, D.S.P.; Martins, M.; Amaral, L. Novel materials for fuel cells operating on liquid fuels. *AIMS Energy* **2017**, *5*, 458–481. [CrossRef]
2. Bai, J.; Liu, D.; Yang, J.; Chen, Y. Nanocatalysts for Electrocatalytic Oxidation of Ethanol. *ChemSusChem* **2019**, *12*, 2117–2132. [CrossRef] [PubMed]
3. Lamy, C.; Belgsir, E.M.; Léger, J.M. Electrocatalytic oxidation of aliphatic alcohols: Application to the direct alcohol fuel cell (DAFC). *J. Appl. Electrochem.* **2001**, *31*, 799–809. [CrossRef]
4. Li, C.; Wang, K.; Xie, D. A review of approaches for the design of high-performance electrocatalysts for ethanol electrooxidation. *Surf. Interfaces* **2022**, *28*, 101594. [CrossRef]
5. Badwal, S.P.S.; Giddey, S.; Kulkarni, A.; Goel, J.; Basu, S. Direct ethanol fuel cells for transport and stationary applications—A comprehensive review. *Appl. Energy* **2015**, *145*, 80–103. [CrossRef]
6. Elsaid, K.; Abdelfatah, S.; Abdel Elabsir, A.M.; Hassiba, R.J.; Ghouri, Z.K.; Vechot, L. Direct alcohol fuel cells: Assessment of the fuel's safety and health aspects. *Int. J. Hydrogen Energy* **2021**, *46*, 30658–30668. [CrossRef]
7. Sanchez, N.; Ruiz, R.; Hacker, V.; Cobo, M. Impact of bioethanol impurities on steam reforming for hydrogen production: A review. *Int. J. Hydrogen Energy* **2020**, *45*, 11923–11942. [CrossRef]
8. Ma, Y.; Ma, C.; Wang, Y.; Wang, K. Advanced Nickel-Based Catalysts for Urea Oxidation Reaction: Challenges and Developments. *Catalysts* **2022**, *12*, 337. [CrossRef]
9. Antolini, E.; Gonzalez, E.R. Alkaline direct alcohol fuel cells. *J. Power Sources* **2010**, *195*, 3431–3450. [CrossRef]
10. Rahim, M.A.A.; Hameed, R.M.A.; Khalil, M.W. Nickel as a catalyst for the electro-oxidation of methanol in alkaline medium. *J. Power Sources* **2004**, *134*, 160–169. [CrossRef]
11. Holade, Y.; Tuleushova, N.; Tingry, S.; Servat, K.; Napporn, T.W.; Guesmi, H.; Cornu, D.; Kokoh, K.B. Recent advances in the electrooxidation of biomass-based organic molecules for energy, chemicals and hydrogen production. *Catal. Sci. Technol.* **2020**, *10*, 3071–3112. [CrossRef]
12. Yuan, L.S.; Zheng, Y.X.; Jia, M.L.; Zhang, S.J.; Wang, X.L.; Peng, C. Nanoporous nickel-copper-phosphorus amorphous alloy film for methanol electro-oxidation in alkaline medium. *Electrochim. Acta* **2015**, *154*, 54–62. [CrossRef]
13. Mansor, M.; Timmiati, S.N.; Lim, K.L.; Wong, W.Y.; Kamarudin, S.K.; Nazirah Kamarudin, N.H. Recent progress of anode catalysts and their support materials for methanol electrooxidation reaction. *Int. J. Hydrogen Energy* **2019**, *44*, 14744–14769. [CrossRef]
14. Radenahmad, N.; Afif, A.; Petra, P.I.; Rahman, S.M.H.; Eriksson, S.G.; Azad, A.K. Proton-conducting electrolytes for direct methanol and direct urea fuel cells—A state-of-the-art review. *Renew. Sustain. Energy Rev.* **2016**, *57*, 1347–1358. [CrossRef]
15. Xu, C.; Wang, X.; Zhu, J. Graphene—Metal particle nanocomposites. *J. Phys. Chem. C* **2008**, *112*, 19841–19845. [CrossRef]
16. Das, S.; Dutta, K.; Shul, Y.G.; Kundu, P.P. Progress in Developments of Inorganic Nanocatalysts for Application in Direct Methanol Fuel Cells. *Crit. Rev. Solid State Mater. Sci.* **2015**, *40*, 316–357. [CrossRef]
17. Bong, S.; Kim, Y.-R.; Kim, I.; Woo, S.; Uhm, S.; Lee, J.; Kim, H. Graphene supported electrocatalysts for methanol oxidation. *Electrochem. Commun.* **2010**, *12*, 129–131. [CrossRef]
18. Askari, M.B.; Salarizadeh, P. Superior catalytic performance of $NiCo_2O_4$ nanorods loaded rGO towards methanol electro-oxidation and hydrogen evolution reaction. *J. Mol. Liq.* **2019**, *291*, 111306. [CrossRef]
19. Noor, T.; Pervaiz, S.; Iqbal, N.; Nasir, H.; Zaman, N.; Sharif, M.; Pervaiz, E. Nanocomposites of NiO/CuO based MOF with rGO: An efficient and robust electrocatalyst for methanol oxidation reaction in DMFC. *Nanomaterials* **2020**, *10*, 1601. [CrossRef]
20. Urbańczyk, E.; Maciej, A.; Simka, W. Electrocatalytic properties of Co decorated graphene and graphene oxide for small organic molecules oxidation. *Int. J. Hydrogen Energy* **2020**, *45*, 1769–1783. [CrossRef]
21. Solmaz, R.; Döner, A.; Kardaş, G. Electrochemical deposition and characterization of NiCu coatings as cathode materials for hydrogen evolution reaction. *Electrochem. Commun.* **2008**, *10*, 1909–1911. [CrossRef]
22. Faid, A.Y.; Barnett, A.O.; Seland, F.; Sunde, S. NiCu mixed metal oxide catalyst for alkaline hydrogen evolution in anion exchange membrane water electrolysis. *Electrochim. Acta* **2021**, *371*, 137837. [CrossRef]
23. Zhang, L.; Zheng, Y.; Wang, J.; Geng, Y.; Zhang, B.; He, J.; Xue, J.; Frauenheim, T.; Li, M. Ni/Mo Bimetallic-Oxide-Derived Heterointerface-Rich Sulfide Nanosheets with Co-Doping for Efficient Alkaline Hydrogen Evolution by Boosting Volmer Reaction. *Small* **2021**, *17*, 2006730. [CrossRef] [PubMed]

24. Lu, X.; Yang, P.; Wan, Y.; Zhang, H.; Xu, H.; Xiao, L.; Li, R.; Li, Y.; Zhang, J.; An, M. Active site engineering toward atomically dispersed M−N−C catalysts for oxygen reduction reaction. *Coord. Chem. Rev.* **2023**, *495*, 215400. [CrossRef]
25. An, Y.; Ijaz, H.; Huang, M.; Qu, J.; Hu, S. The one-pot synthesis of CuNi nanoparticles with a Ni-rich surface for the electrocatalytic methanol oxidation reaction. *Dalton Trans.* **2020**, *49*, 1646–1651. [CrossRef]
26. Abutaleb, A. Electrochemical oxidation of urea on NiCu alloy nanoparticles decorated carbon nanofibers. *Catalysts* **2019**, *9*, 397. [CrossRef]
27. Hefnawy, M.A.; Fadlallah, S.A.; El-Sherif, R.M.; Medany, S.S. Synergistic effect of Cu-doped NiO for enhancing urea electrooxidation: Comparative electrochemical and DFT studies. *J. Alloys Compd.* **2022**, *896*, 162857. [CrossRef]
28. Zhang, L.; Wang, Y.; Liu, Z.; Yan, Z.; Zhu, L. Bimetallic nickel and copper supported pt catalyst for ethanol electro-oxidation in alkaline solution. *Int. J. Electrochem. Sci.* **2018**, *13*, 2164–2174. [CrossRef]
29. Patel, D.A.; Giannakakis, G.; Yan, G.; Ngan, H.T.; Yu, P.; Hannagan, R.T.; Kress, P.L.; Shan, J.; Deshlahra, P.; Sautet, P.; et al. Mechanistic Insights into Nonoxidative Ethanol Dehydrogenation on NiCu Single-Atom Alloys. *ACS Catal.* **2023**, *13*, 4290–4303. [CrossRef]
30. De Waele, J.; Galvita, V.V.; Poelman, H.; Gabrovska, M.; Nikolova, D.; Damyanova, S.; Thybaut, J.W. Ethanol dehydrogenation over Cu catalysts promoted with Ni: Stability control. *Appl. Catal. A Gen.* **2020**, *591*, 117401. [CrossRef]
31. Jafarian, M.; Moghaddam, R.B.; Mahjani, M.G.; Gobal, F. Electro-catalytic oxidation of methanol on a Ni-Cu alloy in alkaline medium. *J. Appl. Electrochem.* **2006**, *36*, 913–918. [CrossRef]
32. Zheng, Y.; Qiao, J.; Yuan, J.; Shen, J.; Wang, A.-j.; Gong, P.; Weng, X.; Niu, L. Three-dimensional NiCu layered double hydroxide nanosheets array on carbon cloth for enhanced oxygen evolution. *Electrochim. Acta* **2018**, *282*, 735–742. [CrossRef]
33. Roy, A.; Talarposhti, M.R.; Normile, S.J.; Zenyuk, I.V.; De Andrade, V.; Artyushkova, K.; Serov, A.; Atanassov, P. Nickel-Copper Supported on Carbon Black Hydrogen Oxidation Catalyst Integrated into Anion-Exchange Membrane Fuel Cell. *Sustain. Energy Fuels* **2018**, *2*, 2268–2275. [CrossRef]
34. Zhao, X.; Fuji, M.; Shirai, T.; Watanabe, H.; Takahashi, M.; Zuo, Y. Electrocatalytic evolution of hydrogen on the NiCu/Al_2O_3/nano-carbon network composite electrode. *J. Mater. Sci.* **2011**, *46*, 4630–4637. [CrossRef]
35. Motlak, M.; Barakat, N.A.M.; El-Deen, A.G.; Hamza, A.M.; Obaid, M.; Yang, O.-B.; Akhtar, M.S.; Khalil, K.A. NiCu bimetallic nanoparticle-decorated graphene as novel and cost-effective counter electrode for dye-sensitized solar cells and electrocatalyst for methanol oxidation. *Appl. Catal. A Gen.* **2015**, *501*, 41–47. [CrossRef]
36. Shoeb, M.; Mobin, M.; Rauf, M.A.; Adnan, S.M.; Ansari, M.Y. Graphene nickel[sbnd]copper nanocomposite (Gr@NiCu NCs) as a binder free electrode for high energy density supercapacitor and antimicrobial application. *J. Mater.* **2021**, *7*, 815–827. [CrossRef]
37. Javan, H.; Asghari, E.; Ashassi-Sorkhabi, H. Design of new anodic bimetallic nanocatalyst composed of Ni–Cu supported by reduced carbon quantum dots for the methanol oxidation reaction. *Diam. Relat. Mater.* **2021**, *115*, 108348. [CrossRef]
38. Rana, S.; Jonnalagadda, S.B. A facile synthesis of Cu-Ni bimetallic nanoparticle supported organo functionalized graphene oxide as a catalyst for selective hydrogenation of p-nitrophenol and cinnamaldehyde. *RSC Adv.* **2017**, *7*, 2869–2879. [CrossRef]
39. Hou, C.; Zhang, M.; Halder, A.; Chi, Q. Graphene directed architecture of fine engineered nanostructures with electrochemical applications. *Electrochim. Acta* **2017**, *242*, 202–218. [CrossRef]
40. Vilana, J.; Gómez, E.; Vallés, E. Influence of the composition and crystalline phase of electrodeposited CoNi films in the preparation of CoNi oxidized surfaces as electrodes for urea electro-oxidation. *Appl. Surf. Sci.* **2016**, *360*, 816–825. [CrossRef]
41. Maniam, K.K.; Chetty, R.; Thimmappa, R.; Paul, S. Progress in the Development of Electrodeposited Catalysts for Direct Liquid Fuel Cell Applications. *Appl. Sci.* **2022**, *12*, 501. [CrossRef]
42. Wala, M.; Szewczyk, M.; Leśniak-Ziółkowska, K.; Kazek-Kęsik, A.; Simka, W. Preparation of NiCuGO composite and investigation of its electrocatalytic properties in methanol oxidation. *Electrochim. Acta* **2022**, *425*, 140743. [CrossRef]
43. Wala, M.; Blacha-Grzechnik, A.; Stolarczyk, A.; Bajkacz, S.; Dydo, P.; Simka, W. Unexpected electrochemical oxidation of urea on a new NiCuGO composite catalyst. *Int. J. Hydrogen Energy* **2023**, *48*, 34229–34243. [CrossRef]
44. Barakat, N.A.M.; Motlak, M. CoxNiy-decorated graphene as novel, stable and super effective non-precious electro-catalyst for methanol oxidation. *Appl. Catal. B Environ.* **2014**, *154–155*, 221–231. [CrossRef]
45. Yang, J.; Shen, X.; Ji, Z.; Zhou, H.; Zhu, G.; Chen, K. In situ growth of hollow CuNi alloy nanoparticles on reduced graphene oxide nanosheets and their magnetic and catalytic properties. *Appl. Surf. Sci.* **2014**, *316*, 575–581. [CrossRef]
46. Glass, D.E.; Galvan, V.; Prakash, G.K.S. The Effect of Annealing Temperature on Nickel on Reduced Graphene Oxide Catalysts on Urea Electrooxidation. *Electrochim. Acta* **2017**, *253*, 489–497. [CrossRef]
47. Liu, X.; Liu, W.; Ko, M.; Park, M.; Kim, M.G.; Oh, P.; Chae, S.; Park, S.; Casimir, A.; Wu, G.; et al. Metal (Ni, Co)-Metal Oxides/Graphene Nanocomposites as Multifunctional Electrocatalysts. *Adv. Funct. Mater.* **2015**, *25*, 5799–5808. [CrossRef]
48. Duan, D.; Gao, J.; Wang, Y.; Zhou, X.; Liu, S.; Wang, Y. Electrodeposition of copper-cobalt bimetallic phosphide on nickel foam as an efficient catalyst for overall water splitting. *J. Electroanal. Chem.* **2023**, *939*, 117478. [CrossRef]
49. Sun, Q.; Li, Y.; Wang, J.; Cao, B.; Yu, Y.; Zhou, C.; Zhang, G.; Wang, Z.; Zhao, C. Pulsed electrodeposition of well-ordered nanoporous Cu-doped Ni arrays promotes high-efficiency overall hydrazine splitting. *J. Mater. Chem. A* **2020**, *8*, 21084–21093. [CrossRef]
50. Lotfi, N.; Shahrabi, T.; Yaghoubinezhad, Y.; Barati Darband, G. Electrodeposition of cedar leaf-like graphene Oxide@Ni–Cu@Ni foam electrode as a highly efficient and ultra-stable catalyst for hydrogen evolution reaction. *Electrochim. Acta* **2019**, *326*, 134949. [CrossRef]

51. Yan, W.; Wang, D.; Diaz, L.A.; Botte, G.G. Nickel nanowires as effective catalysts for urea electro-oxidation. *Electrochim. Acta* **2014**, *134*, 266–271. [CrossRef]
52. Plowman, B.J.; Jones, L.A.; Bhargava, S.K. Building with bubbles: The formation of high surface area honeycomb-like films via hydrogen bubble templated electrodeposition. *Chem. Commun.* **2015**, *51*, 4331–4346. [CrossRef] [PubMed]
53. Yang, S.B.; Tsai, Y.C.; Wu, M.S. Honeycomb-like copper/cuprous oxide with supported nickel hydroxide layer as an electrode material for electrochemical oxidation of urea. *J. Alloys Compd.* **2020**, *836*, 155533. [CrossRef]
54. Zhang, H.; Ye, Y.; Shen, R.; Ru, C.; Hu, Y. Effect of Bubble Behavior on the Morphology of Foamed Porous Copper Prepared via Electrodeposition. *J. Electrochem. Soc.* **2013**, *160*, D441–D445. [CrossRef]
55. Shin, H.C.; Liu, M. Copper foam structures with highly porous nanostructured walls. *Chem. Mater.* **2004**, *16*, 5460–5464. [CrossRef]
56. Li, R.; Li, Y.; Yang, P.; Ren, P.; Wang, D.; Lu, X.; Zhang, H.; Zhang, Z.; Yan, P.; Zhang, J.; et al. Key Roles of Interfacial OH^- ion Distribution on Proton Coupled Electron Transfer Kinetics toward Urea Oxidation Reaction. *Small* **2023**, *19*, 2302151. [CrossRef] [PubMed]
57. Deng, Z.; Yi, Q.; Zhang, Y.; Nie, H. NiCo/C-N/CNT composite catalysts for electro-catalytic oxidation of methanol and ethanol. *J. Electroanal. Chem.* **2017**, *803*, 95–103. [CrossRef]
58. Miao, Z.; Xu, C.; Zhan, J.; Xu, Z. Morphology-control and template-free fabrication of bimetallic Cu–Ni alloy rods for ethanol electro-oxidation in alkaline media. *J. Alloys Compd.* **2021**, *855*, 157438. [CrossRef]
59. Nassif, N.; Ghayad, I.; Hamid, Z.A. Review article: Direct ethanol fuel cells as a renewable source of energy. *Egypt. J. Chem.* **2021**, *64*, 2723–2729. [CrossRef]
60. Martín-Yerga, D.; Henriksson, G.; Cornell, A. Insights on the ethanol oxidation reaction at electrodeposited PdNi catalysts under conditions of increased mass transport. *Int. J. Hydrogen Energy* **2021**, *46*, 1615–1626. [CrossRef]
61. Akhairi, M.A.F.; Kamarudin, S.K. Catalysts in direct ethanol fuel cell (DEFC): An overview. *Int. J. Hydrogen Energy* **2016**, *41*, 4214–4228. [CrossRef]
62. Almeida, C.V.S.; Tremiliosi-Filho, G.; Eguiluz, K.I.B.; Salazar-Banda, G.R. Improved ethanol electro-oxidation at Ni@Pd/C and Ni@PdRh/C core–shell catalysts. *J. Catal.* **2020**, *391*, 175–189. [CrossRef]
63. Altarawneh, R.M.; Brueckner, T.M.; Chen, B.; Pickup, P.G. Product distributions and efficiencies for ethanol oxidation at PtNi octahedra. *J. Power Sources* **2018**, *400*, 369–376. [CrossRef]
64. Guchhait, S.K.; Paul, S. Electrochemical development of Ni-Cu electrodes by direct and pulse current coating in ethanol electro-oxidation for DEFC. *Port. Electrochim. Acta* **2018**, *36*, 293–307. [CrossRef]
65. Rahmani, K.; Habibi, B. NiCo alloy nanoparticles electrodeposited on an electrochemically reduced nitrogen-doped graphene oxide/carbon-ceramic electrode: A low cost electrocatalyst towards methanol and ethanol oxidation. *RSC Adv.* **2019**, *9*, 34050–34064. [CrossRef] [PubMed]
66. Ghalkhani, M.; Mirzaie, R.A.; Banimostafa, A.; Sohouli, E.; Hashemi, E. Electrosynthesis of ternary nonprecious Ni, Cu, Fe oxide nanostructure as efficient electrocatalyst for ethanol electro-oxidation: Design strategy and electrochemical performance. *Int. J. Hydrogen Energy* **2023**, *48*, 21214–21223. [CrossRef]
67. Yu, J.; Ni, Y.; Zhai, M. Simple solution-combustion synthesis of Ni-NiO@C nanocomposites with highly electrocatalytic activity for methanol oxidation. *J. Phys. Chem. Solids* **2018**, *112*, 119–126. [CrossRef]
68. Cai, J.; Zeng, Y.; Guo, Y. Copper@palladium-copper core-shell nanospheres as a highly effective electrocatalyst for ethanol electro-oxidation in alkaline media. *J. Power Sources* **2014**, *270*, 257–261. [CrossRef]
69. Zheng, Y.; Wan, X.; Cheng, X.; Cheng, K.; Liu, Z.; Dai, Z. Advanced catalytic materials for ethanol oxidation in direct ethanol fuel cells. *Catalysts* **2020**, *10*, 166. [CrossRef]
70. Hassan, H.B.; Hamid, Z.A. Electrodeposited Ni-Cr_2O_3 nanocomposite anodes for ethanol electrooxidation. *Int. J. Hydrogen Energy* **2011**, *36*, 5117–5127. [CrossRef]
71. Lycke, D.R.; Gyenge, E.L. Electrochemically assisted organosol method for Pt-Sn nanoparticle synthesis and in situ deposition on graphite felt support: Extended reaction zone anodes for direct ethanol fuel cells. *Electrochim. Acta* **2007**, *52*, 4287–4298. [CrossRef]
72. Bender, M.T.; Lam, Y.C.; Hammes-Schiffer, S.; Choi, K.S. Unraveling Two Pathways for Electrochemical Alcohol and Aldehyde Oxidation on NiOOH. *J. Am. Chem. Soc.* **2020**, *142*, 21538–21547. [CrossRef] [PubMed]
73. Kim, J.-W.; Park, S.-M. Electrochemical Oxidation of Ethanol at Nickel Hydroxide Electrodes in Alkaline Media Studied by Electrochemical Impedance Spectroscopy. *J. Korean Electrochem. Soc.* **2005**, *8*, 117–124. [CrossRef]
74. Chemchoub, S.; El Attar, A.; Oularbi, L.; Younssi, S.A.; Bentiss, F.; Jama, C.; El Rhazi, M. Electrosynthesis of eco-friendly electrocatalyst based nickel-copper bimetallic nanoparticles supported on poly-phenylenediamine with highest current density and early ethanol oxidation onset potential. *Int. J. Hydrogen Energy* **2022**, *47*, 39081–39096. [CrossRef]
75. Kamyabi, M.A.; Jadali, S. Rational design of PdCu nanoparticles supported on a templated Ni foam: The cooperation effect of morphology and composition for electrocatalytic oxidation of ethanol. *Int. J. Hydrogen Energy* **2021**, *46*, 39387–39403. [CrossRef]

energies MDPI

Article

Magnesium–Air Battery with Increased Power Using Commercial Alloy Anodes

Andrey Zhuk, Grigory Belyaev, Tatiana Borodina, Elena Kiseleva *, Eugeny Shkolnikov, Viktor Tuganov, Georgy Valiano and Viktor Zakharov

Joint Institute for High Temperatures of the Russian Academy of Sciences, Izhorskayast. 13 Bd.2, 125412 Moscow, Russia; zakharovvp@yandex.ru (V.Z.)
* Correspondence: kanna787@mail.ru

Abstract: Mg–air batteries have high theoretical energy density and cell voltage. Their use of environmentally friendly salt electrolyte and commercially available magnesium materials determines their acceptable technical and economic efficiency, safety, and ease of operation. However, the practical applicationsof Mg–air batteries arevery limited due to the polarization of magnesium anodes and the batteries' low Faraday efficiency. In this study, we considered the possibility of designingan Mg–air battery withincreased power by adapting engineering solutions developed for an Al–air battery with alkaline electrolytes. To increase the specific power of the battery, it was proposed that the internal resistance of the battery maybe reduced using a concentrated salt electrolyte. We investigated the discharge performance of a commercial alloy of AZ31 type in 15 wt.% NaCl electrolyte at current densities of 40–120 mA/cm^2. The influence of a small addition of sulfosalicylic acid into the electrolyte on the discharge performance of the anode alloy was studied as well. The estimated values of the energy characteristics of the 0.5 kW Mg–air battery were compared with those of an Al–air battery with an alkaline electrolyte.

Keywords: clean energy; metal energy carriers; metal–air battery; Mg–air battery; fuel cell

1. Introduction

Recently, Li–ion and Pb–acid batteries have beenthe most popular types of secondary electrochemical power sources. Usually, stationary electrical circuits are used to charge these batteries. For the emergency charging of EV batteries or charging of batteries operated in field conditions, mobile charging stations with electric generators based on internal combustion engines (ICEs) are used. Such charging stations require periodic maintenance. Their operation isaccompanied by greenhouse gas emissions. An alternative to ICE-based mobile charging stations ismetal–air primary electrochemical generators (batteries).

Metal–air batteries (Al–air, Zn–air, and Mg–air) can be successfully used as emergency and backup power sources as well as chargers for secondary batteries. Metal–air batteries are recharged by replacing the anodes. These batteries are started (activated) by pouring liquid electrolyte into them. Their main advantages are high specific energy (up to 300–500 Wh/kg) and an almost unlimited shelf life in an unactivated state. In anunactivated state, such batteries do not require maintenance.

Mg–air batteries have high theoretical energy density and cell voltage. The use of environmentally friendly salt electrolyte in combination with available magnesium alloys determines the acceptable technical and economic efficiency, safety, and ease of operation of Mg–air batteries when compared with Al–air and Zn–air analogs.

On the other hand, commercially available Mg-based anodes suffer from problems involvinghigh self-discharge rates and the formation of passive films blocking the discharge activity. Thisis why Mg–air batteries are usually considered less attractive than Al–air and Zn–air batteries, although the latter use aggressive alkaline electrolytes and expensive specialized anode alloys [1,2].

Earlier, we developed an Al–air battery with an alkaline electrolyte and anodes made from99.99 wt.% Al alloyed with 0.4 wt.% In [1], gas diffusion cathodes with activated carbon catalysts that hadacceptable catalytic activity at low cost were used. A 1.5 kW battery has a capacity of about 14 kWh.The average current density overa discharge time of 10.8 h was 150 mA/cm^2.

In this study, we explored the possibility of adapting the previously developed Al–air battery to work with available and environmentally friendly materials—commercial magnesium alloy anodes and NaCl electrolyte. It was assumed that the Mg–air battery should have enough power to be used as an emergency and/or backup source as well as a charger for Pb–acid and Li–ion batteries with a capacity of about 50–100 Ah.

Recently, a significant number of studies have been devoted to the improvement of the discharge characteristics of Mg anodes. Very informative, comprehensive reviews of these studies are presented in [3–5]. To increase the efficiency of magnesium anodes, various alloying elements (Al, Li, Pb, Sn, Ca, Ga, In, and rare earth elements) are applied, sometimes in combination with severe plastic deformation. It has beenshown that Mg alloys with refined microstructure as well as fine and uniform precipitates are prospectivematerials for high-performance anodes [6,7]. In most cases, studies were carried out at low (1–10 mA/cm^2) current densities. High anodic efficiency (more than 80%) and specific capacity (up to 2100 mAh/g) have been achieved for Mg–Li–Al and Mg–Ca–In alloy anodes [8,9].

The results obtained at low current densities are useful for the development of low-current power sources. For the development of more powerful Mg–air batteries, these results cannot be applied directly. At high current densities, the anode potential decreases significantly, and the corrosion rate increases (a negative differential effect takes place). The formation kinetics of a passivating film on the anode surface may change significantly. An increase incurrent density may be also by accompaniedby an increase involtage loss in the electrolyte viacathode polarization.

Significantly fewerstudies have beendevoted to the investigation of the discharge characteristics of magnesium anodes and Mg–air batteries at higher current densities—more than 30–40 mA/cm^2. Impressive results were obtained for AP65 (Mg–Al–Pb) series alloys [10–12]. At a current density of 150 mA/cm^2 and a Faraday efficiency of 81.7%, a specific power of about 80 mW/cm^2 was achieved. The additions of cerium and lanthanum to the starting AP 65 alloy allowed the specific power to increase up to 90–100 mW/cm^2. Unfortunately, the presence of lead in Mg–Al–Pb alloys makes the handling of reaction products potentially hazardous to the environment.

Alloys of the AT series (Mg–Al–Sn) also allow for high (up to 150 mA/cm^2) current densities to be obtained. After additional thermal or mechanical treatment, the AT61 cast alloy allowed for a maximum specific power of about 90 mW/cm^2 to be obtained at a Faraday efficiency of more than 70% [13]. It should be noted that without thermomechanical treatment, the specific power and Faraday efficiency of similar cast alloys appeared to be noticeably lower [14].

AMg–Li–Al–Ce–Y–Zn anode alloy was studied in ref. [15].The authors used a commercial alloy, AZ31, as theirstarting material. Apeak power density of 92 mW/cm^2 at a Faraday efficiency of 75% was obtained.

In ref. [16], theanode alloy Mg-1Bi-0.5Ca was studied. The maximum power was 86 mW/cm^2 at a Faraday efficiency of about 75%. The ref. [17] was devoted to the study of extruded alloys of the TZQ (Mg–Sn–Zn–Ag) series. The maximum specific power was about 90 mA/cm^2 with a Faraday efficiency of about 70%.

One may conclude that the additionally alloyed Mg of LA/LAZ, AP, AT, AZ and TZQ series hasquite satisfactory discharge performances, allowing us toobtain a maximum specific power of 80–90 mW/cm^2 with a Faraday efficiency of more than 70%. It is usually assumed that the positive effect is achieved by the formation of intermetallic micro-inclusions, which increase the electrochemical activity without significantly reducing the corrosion resistance of the anode material [4]. However, it should be noted that the practical application of the specialized, high-cost anode alloys is limited until their production reaches the commercial stage. This means that when using specialized alloys, one of the main advantages of Mg–air batteries—the availability and relative cheapness of the anode material—is disappearing.

Commercial Mg alloys are attractive due to their availability and predictable composition. In particular, a significant number of articles are devoted to the study of the AZ series alloys (Mg–Al–Zn) AZ31, AZ61, and AZ91.Currently, these alloys are used by many authors for comparison with new, more efficient anode materials. Important advantages of the AZ series alloys are their availability, relative cheapness, and good discharge performance. For example, the AZ31 alloy allows us to obtain a maximum power of 50–60 mW/cm^2 at a utilization efficiency of 60–65% [14,15]. This alloy is produced commercially in the form of sheets with a thickness of several mm and can successfully compete with specialized anode alloys, especially if its discharge performances can be improved by at least 10–15%.

The second approach used to improve Mg–air batteries characteristics is the selection of an appropriate electrolyte [18,19]. It has been shown that by tuning the pH value, one can suppress hydrogen evolution. In experiments, an electrolyte consisting of a near-saturated aqueous solution of LiCl and $MgCl_2$ was used. The results were obtained at very low (first tens of microamperes) current densities. Also, it was found that the interaction of Cl^- and SO_4^{2-} ions with Mg leads to high corrosion rates, while NO_3^- or NO_2^- ions do not promote corrosion to the same extent as chlorides. The reason for the high corrosion resistance of Mg in the alkaline solution is due to the formation of $Mg(OH)_2$ film on the surface of Mg alloys. However, thick $Mg(OH)_2$ film slows down the reaction of the anode, leading to a decrease in anode potential. Therefore, one may conclude that the applicationsof LiCl, $MgCl_2$, and $Mg(NO_3)_2$ electrolytes do not provide any practical advantages.

Recently, 3.5 wt.% NaCl aqueous electrolyte has been widely used in Mg–air batteries for scientific research as well as for commercial applications.It is proposed that the adoption of seawater as the electrolyte significantly decreases the battery weight, thereby contributing to a high energy density; Surely, if sea-water is available.

Since the adoption of traditional corrosion inhibitors as additives to NaCl electrolyte has not induced any significant improvement in battery performance, it was proposed to use strong Mg^{2+}/Fe^{3+} complexing agents as additives to 3.5 wt.% NaCl electrolyte in order to improve battery performance [20–22]. In particular, it was shown that a cell voltage of 1.85 V and a maximum energy density of 3.0 kWh/kg can be obtained using Mg–Ca microalloyed anodes and NaCl electrolyte with additives of complexing agents [22].

It should be emphasized that the excellent discharge characteristics of the proposed combinations of electrolyte and anode material were obtained at low current densities (less than 10 mA/cm^2).

Much less information in the literature concerns the behavior of Mg anodes in concentrated NaCl solutions [4,5]. At the same time, there is reason to believe that the use of concentrated (~15 wt.% NaCl) electrolyte can provide a significant increase in the specific power of the Mg–air batteries. First, the dependence of the electrical resistance of the NaCl solution on the concentration has a minimum at 15 wt.%. Secondly, chloride ions are activators of magnesium anode dissolution. In this regard, an increase in the concentration of chloride ions contributes to the loosening of the oxide film on the anode surface, i.e., freeing the surface and accelerating the anode processes. However, such an acceleration may lead to an increase in the corrosion rate, which is typical for a negative difference effect [4].

Surprisingly, there is a lack of information on the discharge performance of commercial Mg alloys in 15 wt.% NaCl electrolyte at current densities higher than 40 mA/cm^2. In ref. [23], the discharge characteristics of commercially pure magnesium in 14 wt.% NaCl electrolyte were studied. The measurements were carried out at a current density up to 160 mA/cm^2. Unfortunately, the authors [23] did not give any data on the anode efficiency.

In ref. [24], the corrosion of AZ31 electrodes in 8 wt.% NaCl solution with different concentrations of $CeN_3O_9{\cdot}6H_2O$ additions was studied. Information about the discharge characteristics of the battery is not presented in the article.

In ref. [25], the discharge characteristics of the AZ31 and AZ61 alloys were studied in electrolytes with concentrations 7–18 wt.% of NaCl. The measurements were carried out up to a current density of 40 mA/cm^2. The Faraday efficiency may be estimated by 60–65%.

Summarizing the above brief review, one may conclude that detailed information on the discharge performances of Mg–air batteries with anelectrolyte of 15 wt.% NaCl and with anodes of the commercial AZ series alloys (Mg–Al–Zn) at currents above 40 mA/cm^2 is not available in the literature.

The aim of this study was the investigation of discharge performances of anodes made of the commercially available alloy MA2-2 (AZ31 type) in a 15 wt.% NaCl electrolyte at current densities up to 120 mA/cm^2. Thisstudy also includes the results for 3.5 and 10 wt.%NaCl electrolytes and 15 wt.% NaCl electrolyte with the addition of the strong Mg^{2+}/Fe^{3+} complexing agentsulfosalicylic acid.

2. Materials and Methods

The magnesium alloy MA2-2 was chosen as the anode material. MA2-2 alloy is a Russian analogue of the well-studied alloy AZ31.It is commercially produced by the All-Russian Institute of Light Alloys (Moscow, Russia). The chemical composition of the MA2-2 alloy was studied using an inductively coupled plasma mass spectrometer (ICP-MS, X-2 Thermo Scientific, Thermo Fisher Scientific, Waltham, MA, USA).The chemical compositions of the alloys MA2-2 and AZ31 are presented in Table 1.

Table 1. Compositions of MA2-2 and AZ31 alloys.

	Mg	Al	Zn	Mn	Fe	Cu	Ni	Si	Ca
MA2-2	95.95–97.5	2–2.5	0.7–1.2	0.15–0.35	0.014	0.00042	0.0006	n/a	Ldl *
AZ31 (ASTM B90) [26]	97	3.1	0.73	0.25	0.005	0.05	n/a	0.02	0.0014

* Less than detection limit; n/a:data are not available.

The polarization and corrosion characteristics of the anode alloy were determined in a three-electrode cell [27] with an electrolyte volume of 75 mL.The anodes had a working area of 8 cm^2. The anode temperature was maintained at 30 °C by means of a thermostat. A polymer film electrically isolated the anode reverse side from the thermostat water. The counter electrode was a disk made of a nickel grid. The anode potential was measured relative to the reference Ag | AgCl electrode, which was connected to the cell through a salt bridge and brought to the anode surface using a Luggin capillary. The working surface of each anode was polished with 1200 grit sandpaper. Before measurements, the anode was immersed in the electrolyte for 5 min.

Measurements were carried out at three concentrations of an aqueous electrolyte: 3.5, 10, and 15 wt.% NaCl, as well as in 15 wt.% NaCl electrolyte with additions of 0.003 and 0.012 M sulfosalicylic acid ($C_7H_6O_6S{\cdot}\ H_2O$). The electrolyte was prepared from chemically pure grade reagents and distilled water.

The equilibrium corrosion potential of the MA2-2 alloy was measured in the potentiodynamic mode in the range from −1.8 to −1.4 V. The sweep rate was 1 mV/s. The anode potential dependences on the current density were determined via linear sweep voltammetry in the current density range of 0–120 mA/cm^2. The sweep rate was 1.25 mA/s. A Potentiostat Autolab PGSTAT302N (Metrohm AG, Ionenstrasse, Herisau, Switzerland) was used for the electrochemical measurements. The rate of hydrogen evolution was measured using an RRG-12 (BLM Synergy, Electrozavodskaya 24, Moscow, RF, Russia) gas flow meter. The electrochemical and gas flow measurements were synchronized.

The corrosion current density was calculated from the hydrogen evolution rate. The corrosion current and Faraday efficiency were calculated from readings of the flow meter using Formulas (1) and (2):

$$jc = \frac{z \cdot F}{S} \cdot \frac{p}{R \cdot T} \cdot \frac{dV}{dt} \quad (1)$$

$$kF = ja/(jc + ja) \quad (2)$$

where *ja*, *jc*—anodic and corrosion current densities, kF—Faraday efficiency, *dV/dt*—hydrogen release rate, *p*—pressure, *T*—temperature, *S*—anode square, and *z*—the number of electrons involved in the reaction; *F*—Faraday constant, *R*—universal gas constant.

To determine the current–voltage characteristics model, Mg–air cells with initial inter-electrode gaps of 5 and 8 mm were used. The gap values were chosen according to the size of the inter-electrode gap in an Al–air battery [1] at the beginning and at the end of discharge. The anode of the model battery was a disk with an area of 8 cm^2 and a thickness of 3.5 mm. The gas diffusion electrode of our own production [1] was used as a cathode. The electrolyte volume in cells with a gap of 5–8 mm was 4 and 6.4 mL, respectively. The duration of each experiment with model cells was 600 s.

Electrolyte pH measurements during the continuous discharge were carried out in a three-electrode cell equipped with an electrolyte circulation system.The electrolyte circulation system included a vessel with the additional volume of an electrolyte (800 mL) and a circulation pump with the capacity of 20 mL/min.

A Smart Lab SE Rigaku (CuKα radiation, with astep of 0.02 degrees; Rigaku Corporation 3-9-12, Matsubara-cho Akishima-shi, Tokyo 196-8666, Japan) was applied for an XRD analysis of reaction products. A phase identification and quantitative phase analysis were carried out using the Rigaku Integrated X-ray powder diffraction Software PDXL (ver.2.0). SEM and EDS analyses were carried out using the scanning electron microscope NOVA NanoSem 650 (Thermo Fisher Scientific, Waltham, MA, USA). Microscopic and qualitative EDS analyses of the sample surfaces were carried out in reflected electrons using a ring detector, which makes it possible to obtain compositional contrast on non-smooth surfaces.

The anode surfaces were studied in two states: (i) with not-washed-off reaction products; (ii) with washed-off reaction products. The reaction products were washed off with a weak solution of a mixture of acetic, citric, and oxalic acids. The concentration of the solution was selected in such a way that no visible interaction of Mg with the solution occurred. To minimize the effects of charging with an electron probe, samples with unwashed reaction products were covered with a ~20–30 nm thick layer of graphite in a JEE-420 vacuum setup. No electrically conductive coating was applied to samples with washed-off reaction products.

3. Results and Discussion

3.1. Electrochemical Measurements

Figure 1 shows the results of potentiodynamic polarization tests for MA2-2 alloy at three electrolyte concentrations—3.5, 10, and 15 wt.% of NaCl. The value of the corrosion potential of MA2-2 in an electrolyte concentration of 3.5 wt.% NaCl for the MA2-2 alloy is close to that obtained earlier for the AZ31 alloy [8,24]. With an increase inthe electrolyte concentration from 3.5 to 15 wt.%, the equilibrium corrosion potential changes from -1.48 to -1.55 V. Therefore, MA2-2 alloy hasthe highest electrochemical activity at the electrolyte concentration 15 wt.% of NaCl. At the same time, the corrosion current increases and, correspondingly, the corrosion resistance of the alloy slightly reduces.

Dependences of the anode potential and corrosion current density on the anode current density at three NaCl concentrations are shown in Figure 1b,c. At current densities less than 60 mA/cm^2, the anode potentials in the 3.5 wt.% NaCl electrolyte have the lowest values. In this case, the corrosion current also has the lowest value. In other words, in 3.5 wt.% NaCl electrolyte, at current densities less than 60 mA/cm^2, the MA2-2 alloy demonstrates the lowest anodic (Figure 1b) and corrosion activity (Figure 1c).

In the region of high anode currents, anode potentials and corrosion currents have close values at all electrolyte concentrations.At all electrolyte concentrations, the corrosion current does not exceed 38 mA/cm^2. In a 15 wt.% NaCl electrolyte, the Faraday efficiency increases from 65.5% at 40 mA/cm^2 to 74% at 100 mA/cm^2.

In Figure 1d,e discharge and specific power curvesfor a model Mg–air cell are shown. The maximum specific power of 70 mW/cm^2 is achieved in a 15 wt.% electrolyte at the current density of about 100 mA/cm^2. At current densities of 80–100 mA/cm^2, the voltage of a cell with 15 wt.% NaCl electrolyte is 0.4 V higher than that at the concentration of 3.5 wt.% NaCl. The specific power is 1.8 times higher.

The Faraday efficiency of known magnesium anode alloys usually does not exceed 50–70%, and the maximum Faraday efficiency achieved so far is about 85% [3–5]. The results of our measurements show that the combination of the MA2-2 alloy and a 15 wt.% aqueous solution of NaCl allows for the achievement of a significant increase in the specific power of a battery while maintaining an acceptable (more than 70%) Faraday efficiency of the anode alloy. The electrolyte with an increased NaCl concentration has low electrical resistance and allows for reduction in the internal resistance of the cell. Due to this circumstance, a specific power more inherent to specialized anode alloys was achieved. Of course, the specific power for the MA2-2 alloy is about 20% lower than that obtained for the best samples of specialized anode alloys—80–90 mW/cm^2 [10–17]. However, this disadvantage is compensated for by the availability and significantly lower cost of commercial alloys of the AZ31 type.

Earlier, the authors of ref. [21] studied the effect of 15 strong complexing compounds on the discharge characteristics of a magnesium anode in an aqueous electrolyte. A highly diluted salt electrolyte (0.5 wt.%) and commercial-purity magnesium were used. According to their results, strong complexing agents are able to effectively activate the magnesium anode and provide a higher anodic potential. The positive effect of the additive was ascribed to the formation of soluble complexes with Mg^{2+} and Fe^{3+} ions. A decrease in the concentration of additives in the electrolyte leads to a higher Faraday efficiency and a slight decrease in the anode potential. Among the considered additives, the most attractive in terms of efficiency, availability, and cost is sulfosalicylic acid—$C_7H_6O_6S \cdot H_2O$.

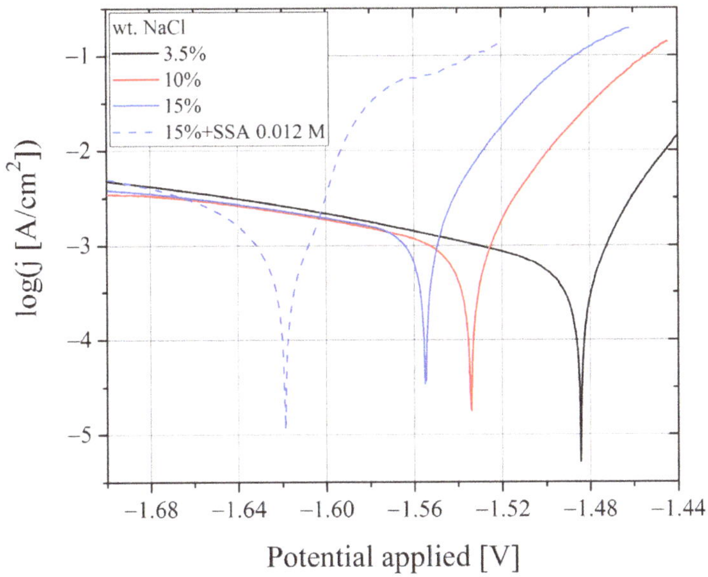

(a)

Figure 1. *Cont.*

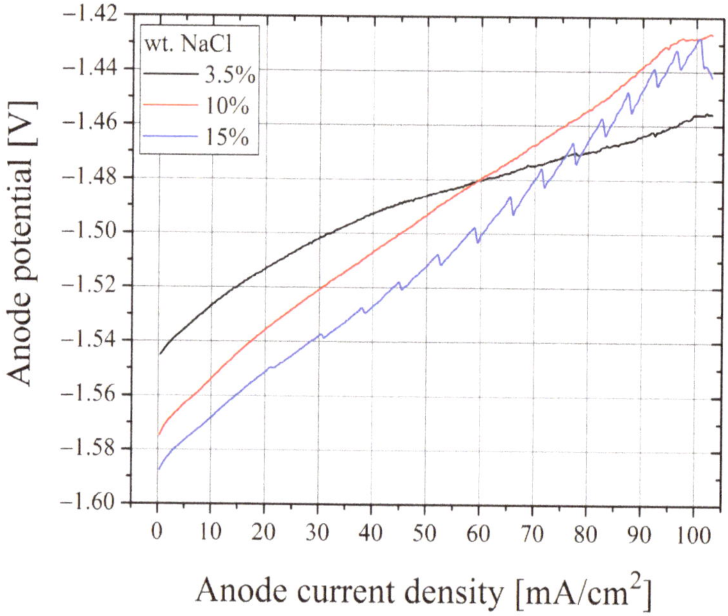

(**b**)

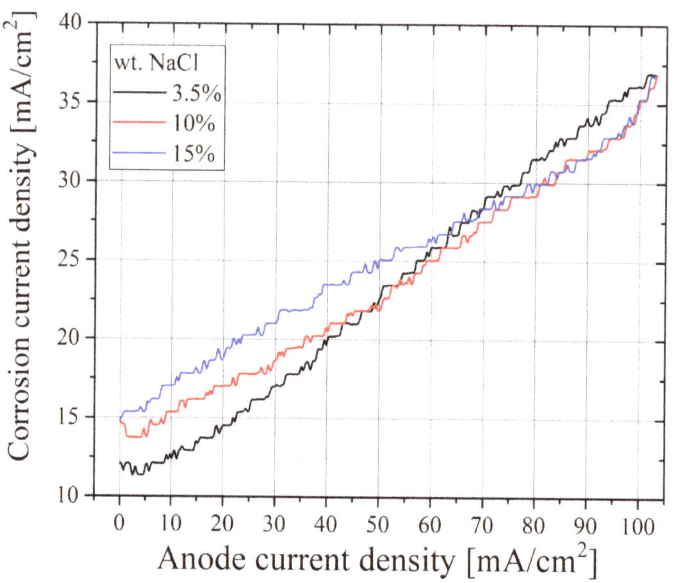

(**c**)

Figure 1. *Cont.*

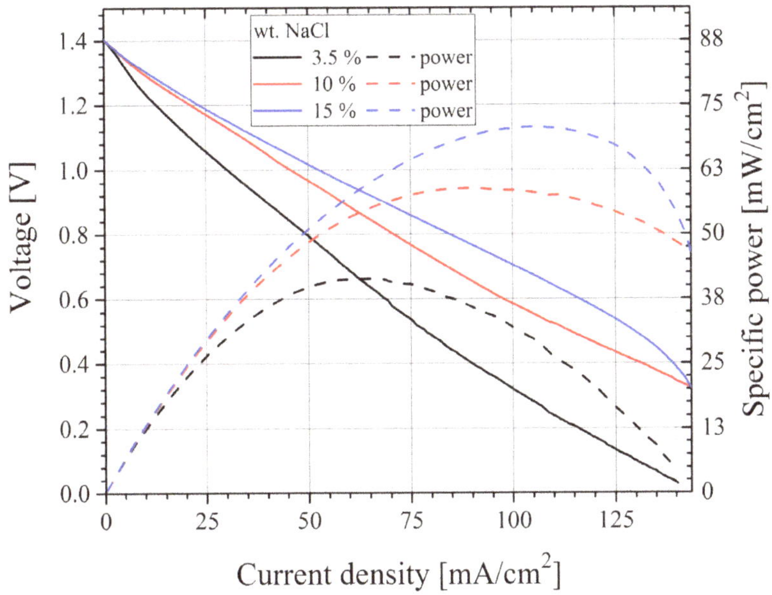

(**d**)

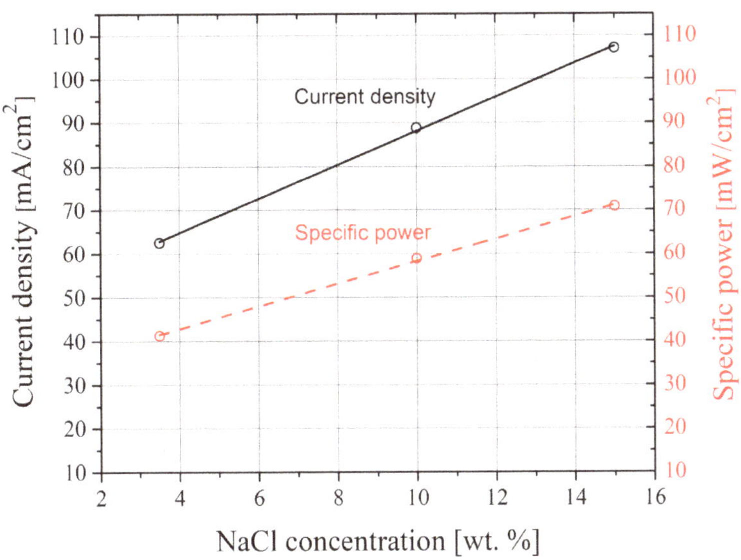

(**e**)

Figure 1. Polarization, corrosion, and battery discharge tests. (**a**) Potentiodynamic polarization tests for MA2-2 alloy at electrolyte concentrations of 3.5, 10, 15 wt.% NaCl, and 15 wt.% NaCl + 0.012 M sulfosalicylic acid. Linear sweep voltammetry tests of MA2-2 alloy at different electrolyte compositions: (**b**) anodic potential vs. anode current density; (**c**) corrosion current density vs. anode current density. Battery tests: (**d**) current–voltage and specific power curves vs. anode current density; (**e**) maximum specific power and current density at the maximum specific power vs. electrolyte concentration.

We studied the effect of adding sulfosalicylic acid (SSA) to a 15 wt.% NaCl solution on the discharge characteristics of the MA2-2 alloy. The experiments were carried out at two concentrations of SSA—0.003 and 0.012 M. It appeared that the addition of SSA does increase the anodic potential at the initial stage of discharge—(Figure 2a,c). This effect occurs approximately up to discharges of 100 C/cm^2. However, as the discharge progresses, this effect disappears and the values of the anodic potentials in electrolytes without SSA and with SSA take close values. At the initial stage of the discharge, the corrosion current also increases and the correlation with anodic potentials changes (Figure 2b,d).Corrosion currents are higher at higher values of anodic potentials.

The results of electrolyte pH and anodic potential measurements in the circulating electrolyte are presented in Figure 3. In the 15 wt.% NaCl electrolyte without a complexing additive, the pH of the electrolyte was close to 11 throughout the experiment. In the 15 wt.% NaCl electrolyte with the addition of 0.012 M SSA, the pH gradually decreased from 1.2–1.6 at the beginning of the discharge and then began to quickly increase up to 7.2 at 300 C/cm^2. A noticeable growth in pH begins at a discharge of 175 C/cm^2.

Up to the discharge of 150 C/cm^2, the anodic potential in an electrolyte with SSA addition is noticeably higher than that in a pure NaCl electrolyte. In the pure NaCl electrolyte, slow potential oscillations began after the discharge of 150 C/cm^2. In the electrolyte with SSA addition, slow potential oscillations began when pH growth started to accelerate at the discharge ~175 C/cm^2.

In Figure 4, photographs of anode surfaces after discharges up to 70 and up to 120 C/cm^2 in the pure 15 wt.% NaCl electrolyte and in the same electrolyte with additions of 0.003 M SSA and 0.012 M SSA are shown. The upper part of the reaction product layer represents a white/grey low dense foam-like material. The bottom, adjacent to the anode layer represents a dark, denser material.

After drying, the upper layer converts to a white powder that can be separated from the anode easily. The bottom layer retains its dark color and can be removed only by etching or by machining of the anode surface.

After a discharge up to 70 C/cm^2, the differences in state of the surface of those anodes that worked in electrolytes without the addition of SSA and those with the addition of SSA are clearly visible. The low-density white/grey hydroxide layer on anodes discharged in an electrolyte with SSA additives is fragmented to a greater extent. After a discharge up to 120 C/cm^2, the differences in the state of the surface of anodes discharged in electrolytes with SSA and without it are almost imperceptible.

To conclude this subsection, we would like to highlight the following results:

- The use of an electrolyte with 15 wt.% of NaCl makes it possible to almost double the specific power of the battery relative to a battery with 3.5 wt.% NaCl electrolyte while maintaining a Faraday efficiency more than 70%;
- The addition of SSA does increase the anodic potential at the initial stage of discharge. As the discharge progresses, this effect disappears and the values of the anodic potentials in electrolytes without SSA and with SSA take close values. In the electrolyte with the addition of 0.012 M SSA, the pH gradually decreased from 1.2–1.6 at the beginning of the discharge, then began to quickly grow up to 7.2 at the end of discharge;
- Photographs of the anode surface after discharges of up to 70 C/cm^2 demonstrate that in the electrolyte with SSA additions, surface films are fragmented to a greater extent than anode surface films in the electrolyte without SSA additions. After a discharge of up to 120 C/cm^2, the differences in the state of the surfaces of anodes discharged in electrolytes with SSA and without it are almost imperceptible.

3.2. Microstructure Examination

The influence of the complexing additive on the phase composition and morphology of the passivating film after the discharge up to of 300 C/cm^2 was studied via X-ray diffractometry and scanning electron microscopy.

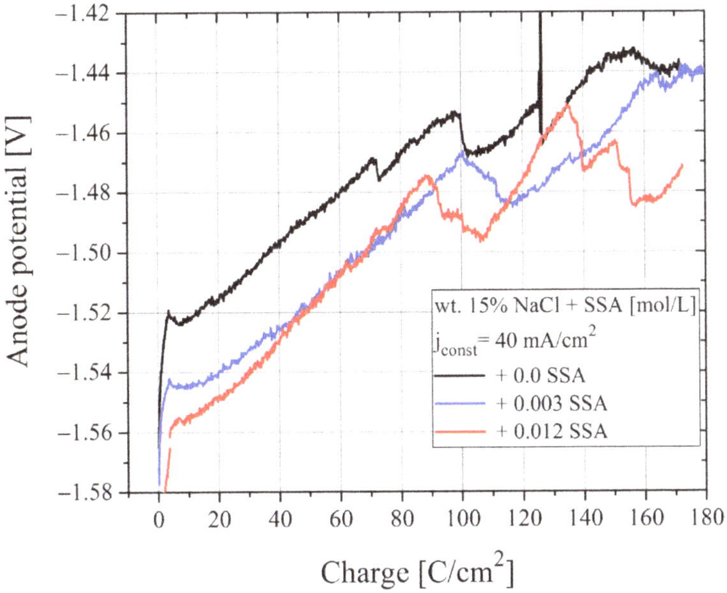

(a)

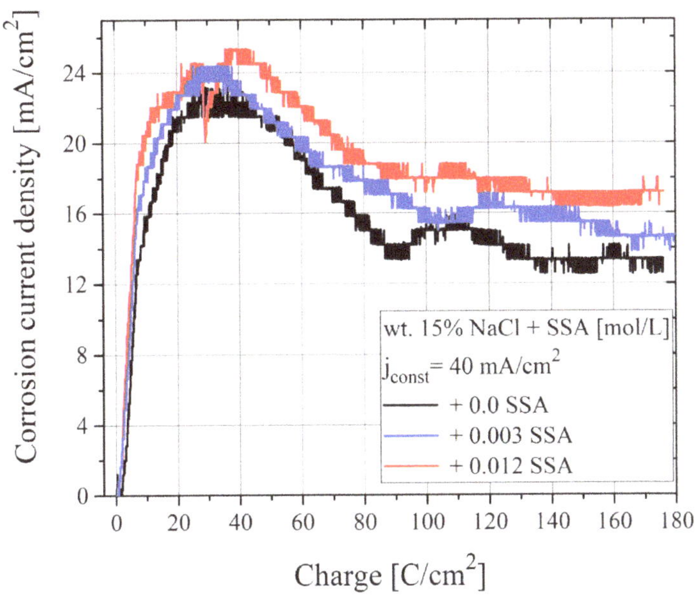

(b)

Figure 2. *Cont.*

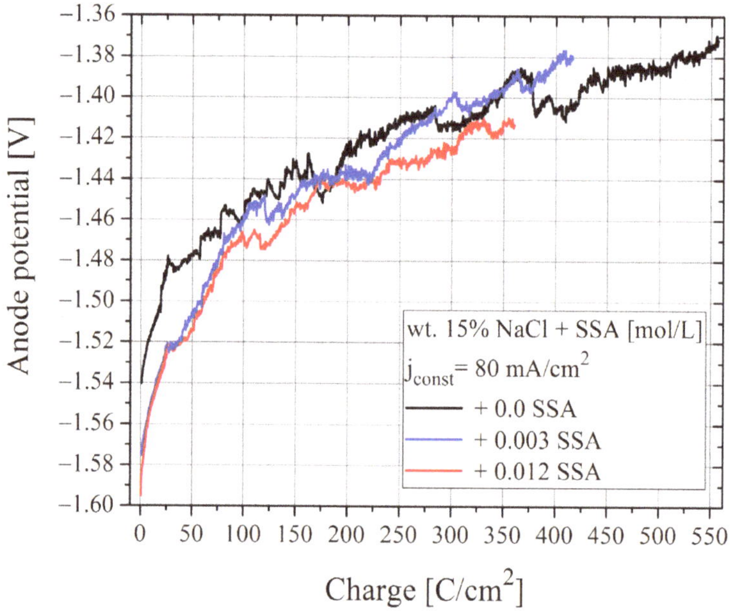

(**c**)

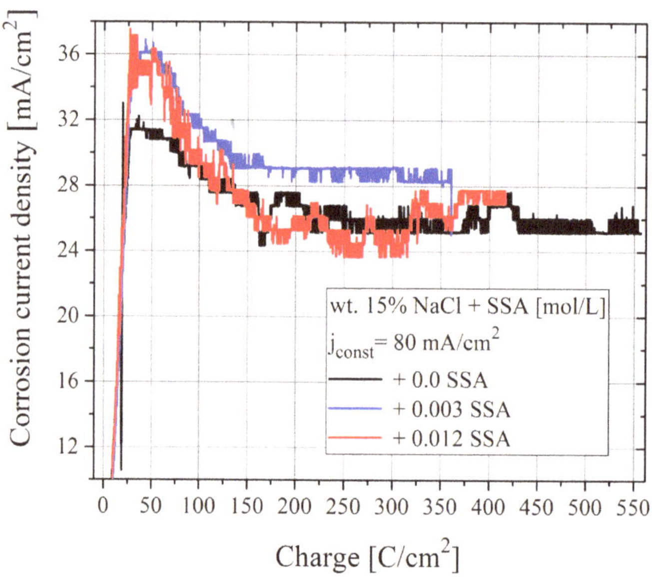

(**d**)

Figure 2. Galvanostatic discharge behavior of MA2-2 alloy at different electrolyte compositions at the anodic current density of 40 mA/cm^2: (**a**) anodic potential; (**b**) corrosion current density; of 80 mA/cm^2: (**c**) anodic potential; (**d**) corrosion current density.

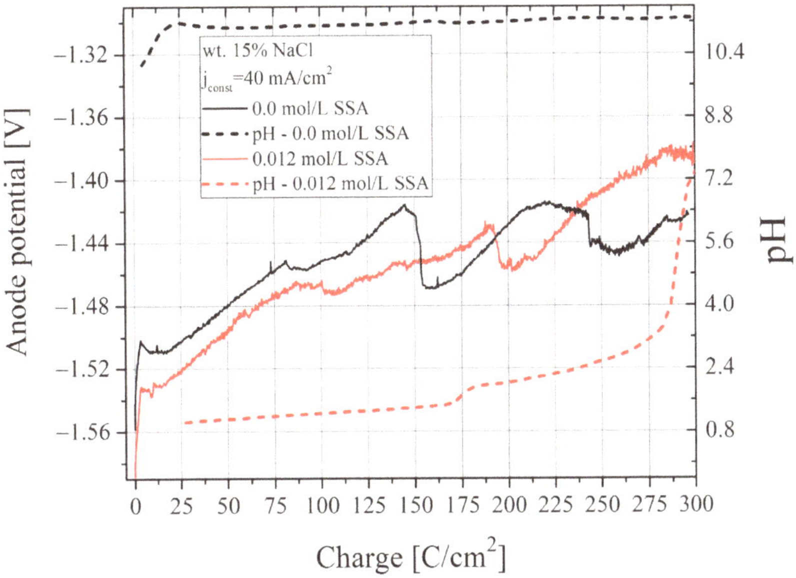

Figure 3. Anodic potential and electrolyte pH measurements during the discharge up to of 300 C/cm^2.

Figure 4. Photos of the anode surfaces after the discharge up to 70 C/cm^2 (**a–c**) and 120 C/cm^2 (**d–f**) in different electrolytes: (**a**,**d**) 15 wt.% NaCl without SSA; (**b**,**e**) 15 wt.% NaCl + 0.003 M SSA; (**c**,**f**) 15 wt.% NaCl + 0.012 M SSA.

The XRD patterns of the alloy oxidation products (passivating film) in electrolytes with SSA addition and without it are practically the same. The material comprises two main phases—magnesium hydroxide $Mg(OH)_2$ (31%) and sodium chloride NaCl (45%). Magnesium hydroxide is represented by two modifications—X-ray amorphous and finely dispersed crystalline. The X-ray amorphous component gives diffuse halos (2θ = 33–42 degree) in the spectrum (Figure 5).

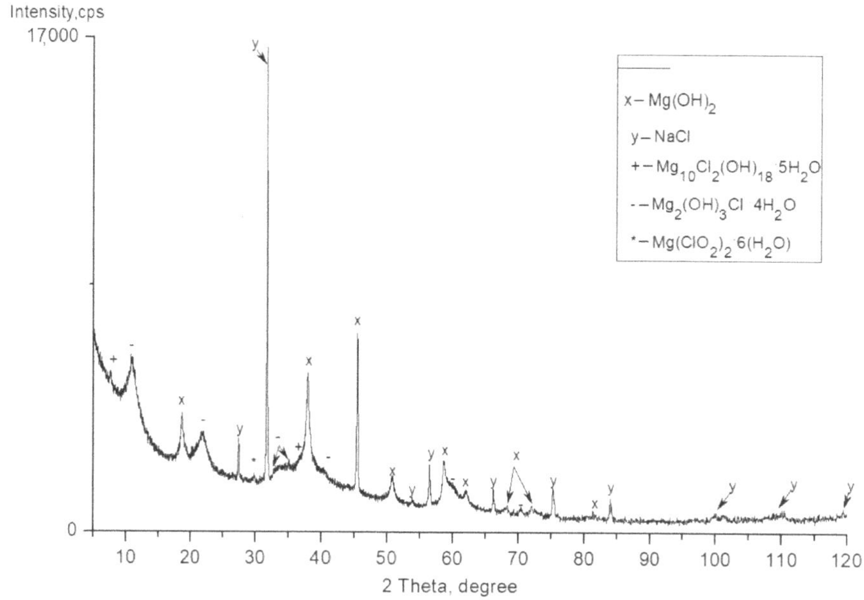

Figure 5. XRDpattern of reaction products with discharge of up to 300 C/cm^2. Diffuse halos between 2θ—33 and 42 degrees.

The size of the coherent scattering region (CSR) of the finely dispersed modification of $Mg(OH)_2$ is 7 nm. Sodium chloride is well crystallized and has an average CSR size of at least 100 nm. The material also includes three phases of magnesium oxychloride cements: $Mg_2(OH)_3Cl{\cdot}4H_2O$ magnesium trihydroxochloridetetrahydrate (22%), $Mg(ClO_2)_2{\cdot}6(H_2O)$ magnesium chlorite hexahydrate (0.5%), and $Mg_{10}Cl_2(OH)_{18}{\cdot}5H_2O$ (1.5%).

It should be emphasized that an addition of 0.012 M SSA leads to a decrease inthe average size of the CSR of $Mg(OH)_2$ and $Mg_2(OH)_3Cl{\cdot}4H_2O$.

The morphology of the anode surface was studied after the discharge in electrolytes with a controlled pH value (Figure 3) up to a discharge of 300 C/cm^2. SEM images of the anode surfaces are presented in (Figure 6).

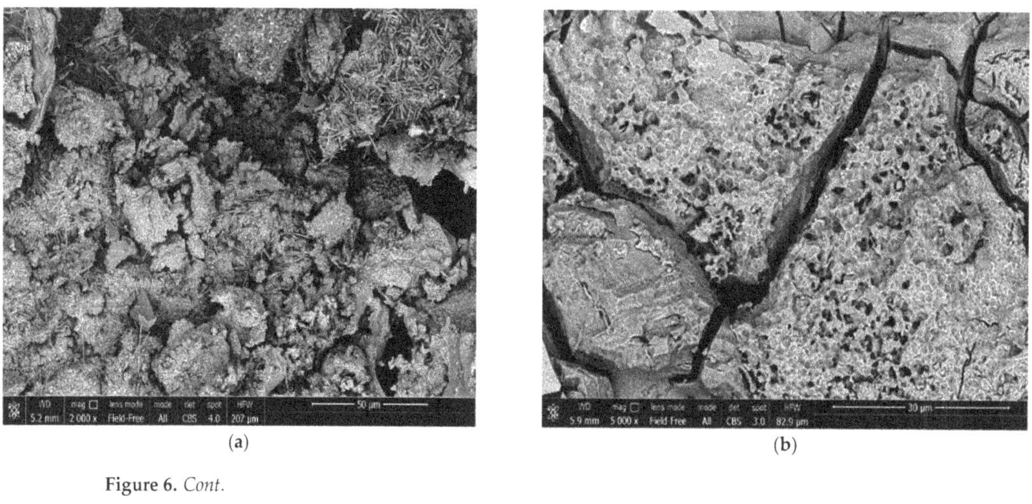

(a) (b)

Figure 6. *Cont.*

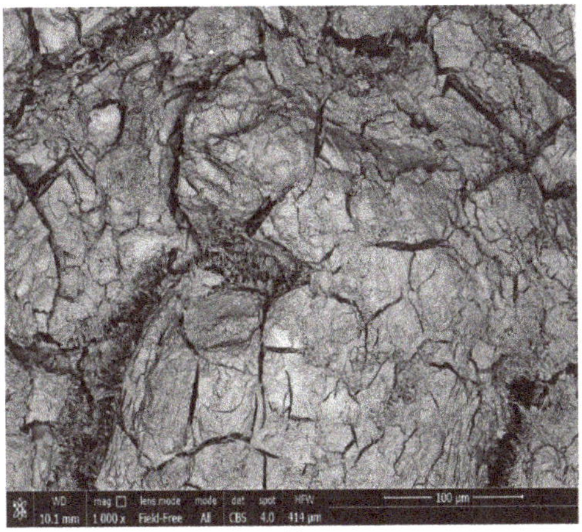

(**c**)

(**d**)

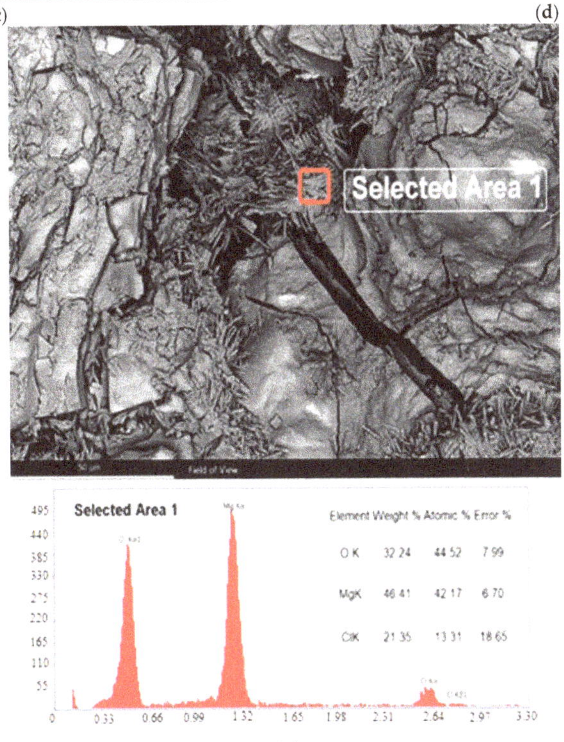

(**e**)

Figure 6. SEM images of the anode surfaces after the discharge test at 40 mA/cm^2 (the discharge up to of 300 C/cm^2) and controlled pH level. (**a**) The anode surface with reaction products, the test in the electrolyte without SSA; (**b**) the anode surface with partially washed-away reaction products, the test in the electrolyte without SSA; (**c**) the anode surface with not-washed-off reaction products, the electrolyte without SSA; (**d**) the anode surface with not-washed-off reaction products, the electrolyte with the addition of 0.012 M SSA; (**e**) EDS analysis of lamella colonies in the cracks.

The anode surface morphologies with not-washed-off reaction products and with washed-off reaction products are presented in (Figure 6a,b), correspondingly. The study shows that the treatment of the anode surface does not allow for the washing off of reaction products completely. However, the removal of a significant amount of fine particles from anode surfaces makes it possible to see corrosion patterns characterized by the presence of crack networks and corrosion pores—(Figure 6b). An EDS analysis captured Mg and oxygen throughout all examined surfaces.

It should be emphasized that after the discharge in the electrolyte without SSA, the anode surface cracks contain significantly more zones with reaction products in the form of lamellae colonies (Figure 6c) than in the samples discharged in the electrolyte with SSA—Figure 6d. An EDS analysis fixed Mg, Cl, and O in these lamellas—(Figure 6e).

To conclude this subsection, we would like to highlight the following results:

- The addition of SSA leads to a decrease in the average size of the CSR of $Mg(OH)_2$ and $Mg_2(OH)_3Cl{\cdot}4H_2O$;
- After anode discharge in the electrolyte without SSA additive, there are noticeably more surface zones with reaction products in the form of lamellae colonies than in the samples discharged in the electrolyte with SSA additive. These lamellas contain Mg, Cl, and O.

3.3. *The Influence of Sulfosalicylic Acid on Mg–Air Battery Performance*

The influence of strong Mg^{2+} complexing agents as additives to 0.5–3.5 NaCl wt.% electrolytes to Mg–air battery performance wasextensively studied by the authors of [20–22,28,29]. It was shown that by adding Mg^{2+}-complexing agents (Tiron, nitrilotriacetic acid, sulfosalicylic acid, etc.) to the aqueous NaCl electrolyte, it was possible to reduce the rate of redepositing of $Mg(OH)_2$ on the anode surface.

In our experiments, the addition of sulfosalicylic acid to the electrolyte leads to an increase in the anode potential accompanied by some decrease in Faraday efficiency. This phenomenon occurs at the initial stage of the discharge (Figure 2a,c). It is clearly seen (Figure 4) that at the initial stage of the discharge in electrolytes with SSA additives, the upper part of hydroxide layers is fragmented to a greater extent. This means that the presence of SSA in the electrolyte slows down the formation of a passivating film on the anode surface.

An SEM examination shows that in the electrolyte without SSA additive, cracks on the anode surface show significantly more zones with reaction products in the form of lamellae colonies (Figure 6c) than on the surface of anodes discharged in the electrolyte with SSA additive (Figure 6d). An EDS analysis fixed Mg, Cl, and O in these lamellas (Figure 6e).

One may propose that the addition of SSA slows down the formation of lamellae colonies on the anode surface. An XRD examination shows that an addition of SSA leads to a decrease in the average size of the CSR of $Mg(OH)_2$ and $Mg_2(OH)_3Cl{\cdot}4H_2O$. Hence, one may propose that the presence of SSA prevents redepositing (formation of lamellae colonies) and slows down the growth of precipitated crystals (which probably make up the lamellae) on the anode surface.

The addition of sulfosalicylic acid to the electrolyte leads to an increase in the anode potential of the alloy at the initial stage of the discharge—Figure 2a,c and Figure 3. When the pH of the electrolyte begins to increase (at discharge 175–190 C/cm^2, see Figure 3), this effect disappears and slow potential oscillations begin to occur. In an electrolyte without SSA additive, slow potential oscillations begin at a discharge of 130–150 C/cm^2.

Earlier, slow potential oscillations of Mg anodes at the deep stage of the discharge were observed by the authors of ref. [30,31]. Following them, we propose that voltage oscillations imply that the active areas of anodes gradually change due to the broken balance between the growth and detachment of reaction products. This is the result of hydrogen release during the corrosion process. Hydrogen accumulates in the cavities inside the passivating film and then breaks through it, partially exposing the anode surface—Figure 4d–f. With increase in the thickness of the passivating film, the volume of the hydrogen accumulation regions also increases. As a result, the frequency of potential oscillations decreases, and their amplitude increases.

Summarizing the above, one may conclude that at the beginning of the discharge, the SSA addition actively interacts with Mg and reaction products on the anode surface. Because of this, the passivating film is formed more slowly than in a pure NaCl electrolyte.As the discharge proceeds, the passivating film thickness gradually increases. Reacting with acid, the passivating film reduces the flow of acid to the anode surface. Adjacent to magnesium, a dense part of the passivating film ceases to dissolve, the anode potential drops, and the corrosion rate decreases. Due to a decrease in the rate of hydrogen evolution and an increase in the thickness of the magnesium hydroxide layer, the rate of the anode self-peeling slows down. The change in the rate of the self-peeling process is observed as the beginning of slow potential oscillations.

Although the presence of SSA slows down the formation of the passivating film, it can be concluded that its addition to the electrolyte does not lead to the improvement of discharge characteristics of the Mg–air battery at later stages of the discharge. Nevertheless, we detected that a small addition of SSA affects the initial state of the anode surface, significantly reducing the dependence of the discharge characteristics on the initial state of the anode—Figure 7a,b. Thus, the addition of sulfosalicylic acid to the electrolyte allows for the stabilization of the operation of the Mg–air battery at the beginning of the discharge without additional treatment of the anode surfaces.

3.4. Comparison of Mg–Airand Al–Air BatteryCharacteristics

In the end, we compared the discharge characteristics of an Al–air battery [1] with an alkaline electrolyte with the estimated characteristics of the Mg–air battery with MA2-2 alloy anodes and 15 wt.% NaCl + 0.003 M SSA electrolyte.

To evaluate the characteristics of the Mg–air battery, the results of the presented study and the geometric dimensions of the Al–air battery were used. It is assumed that the magnesium anodes were plates 100 × 200 × 3.7 mm.

At the current density of 80 mA/cm^2, the Faraday efficiency of the MA2-2 alloy is 75%—Figure 2d. Then, the battery charge, taking into account the complete dissolution of the anodes, will be 210 Ah. At the discharge current density of 80 mA/cm^2, the average voltage during the full discharge time is estimated as 14 V. The average power is 450 W.The peak power at the beginning of the discharge is 630 W at 16.5 V.

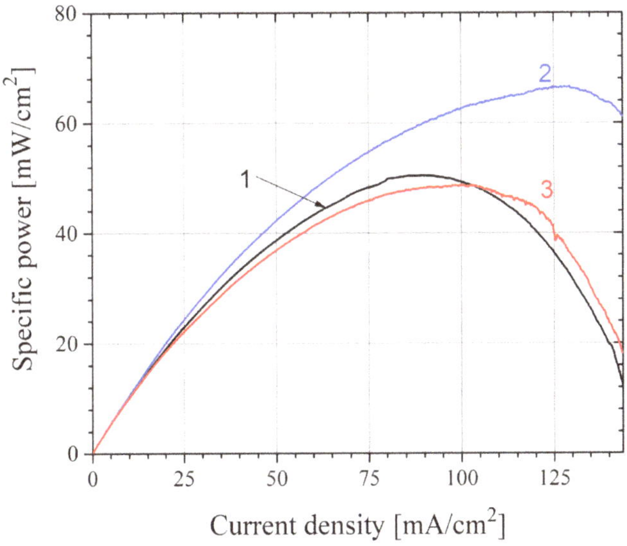

(a)

Figure 7. *Cont.*

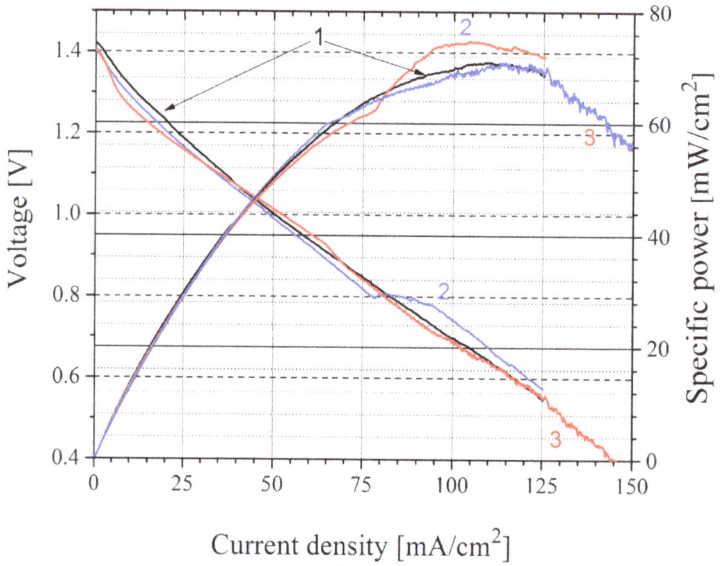

(b)

Figure 7. Battery tests. (**a**) Electrolyte—15 wt. % NaCl. Specific power vs. current density at different initial states of the anodes; (**b**) electrolyte—15 wt. % NaCl + 0.003 M SSA. Specific power vs. current density at different initial states of the anodes. 1—a new anode (polished with 1200 grit sandpaper); 2—the second measurement with the same anode after the battery was short-circuited for 10 s; 3—the same anode after washing the surface with distilled water.

To estimate the electrolyte consumption, experiments witha model single cell with an initial interelectrode gap of 8 mm at a current density of 90 mA/cm^2 were performed. During the discharge, the cell voltage gradually decreased from 0.75 to 0.6 V for 1.2 h and then dropped to zero within 17 min. An estimation of the electrolyte consumption by the time of the beginning of the rapid voltage drop gave 130 Ah/L. Table 2 compares the performance of Al–air and Mg–air batteries.

Table 2. The performance of Al–air and Mg–air batteries.

Characteristics	**Al–Air Battery**	**Mg–Air Battery**
Anode material	Al 99.995 + 0.4 In	Mg MA2-2
Anode weight, kg	4.4	2.83
Water electrolyte composition	32 wt.% NaOH + 0.015 M $Na_2SnO_3 \cdot 3H_2O$	15 wt.% NaCl + 0.003 M $C_7H_6O_6S \cdot H_2O$
Electrolyte volume, L	26	35
Capacity, Ah	580	210
Average electric power, kW	1.4	0.43
Voltage, V	24	14
Current, A	60	40

As expected, in terms of specific energy characteristics, Mg–air batteries are significantly inferior to Al–air analogues. However, taking into account the safety of operation and the availability of consumables, Mg–air batteries can seriously compete with Al–air and Zn–air batteries. For instance, for use as a stationary emergency power source or as a source for battery charging, the loss in specific energy characteristics is unlikely to be critically important.

It is necessary to note one more circumstance concerning Al–air batteries with alkaline (NaOH, KOH) electrolytes. The carbon dioxide contained in the air, joining together with oxygen into the active layer of the gas diffusion cathode, interacts with an aqueous solution of alkali. In this case, salts (Na_2CO_3, K_2CO_3) are formed inside the active layer. These salts have low solubility and can accumulate on the catalyst surface. Salt crystals may reach significant sizes (up to 2–3 mm) and damage cathodes, reducing their life cycle. The carbonation of cathodes does not occur in a salt electrolyte.

4. Conclusions

The discharge and corrosion characteristics of the commercial Mg alloy MA2-2 (Mg–Al–Zn) in 3.5, 10, and 15 wt.% NaCl aqueous solution were studied. It was shown that the use of an electrolyte with 15 wt.% NaCl makes it possible to almost double the specific power of the battery relative to a battery with 3.5 wt.% NaCl electrolyte. The increase in the specific power is achieved due to reducing the resistance of the electrolyte. The Faraday efficiency of the MA2-2 alloy at current densities of 60–100 mA/cm^2 appeared to be more than 70%.

Sulfosalicylic acid, a Mg^{2+} complexing agent, was used as an electrolyte additive. The effect of the electrolyte additive on the corrosion behavior and discharge performance of MA2-2 alloy at current densities up to 100 mA/cm^2 was studied. The selected additive reduces the rate of formation of a passivating film on the anode surface, providing a higher anode activity at the initial stage of the discharge. Sulfosalicylic acid's addition to the electrolyte does not lead to the improvement of the discharge performance of Mg–air battery at a large depth of discharge. However, a small addition of SSA significantly reduces the dependence of battery discharge characteristics on the initial state of the anode surface.

A comparison of the energy performances of an Al–air battery with an alkaline electrolyte and a Mg–air battery with salt electrolyte was conducted. The energy performance of the Mg–air batteries is significantly inferior to their Al–air analogs. However, due to the safety, relative cheapness, and availability of consumables, the use of Mg–air batteries for a number of applications may be preferable.

Author Contributions: Supervision, A.Z.; conceptualization, A.Z.; methodology, A.Z. and E.S.; validation, T.B.; formal analysis, T.B. and E.K.; investigation, G.B., E.K., V.T. and V.Z.; writing—original draft preparation, A.Z., T.B. and G.V.; writing and editing, A.Z.; visualization, G.B. and E.K.; project administration, A.Z. All authors have read and agreed to the published version of the manuscript.

Funding: This work was supported by the Ministry of Science and Higher Education of the Russian Federation (State Assignment No. 075-00270-24-00).

Data Availability Statement: All main experimental data, methods, and materials descriptions are presented in the text of the article. The raw data (tables, XRD spectra, SEM results) are available on request from the first author: Andrey Z. Zhuk, Doctor of Science, Head of Alternative Energy Department, Joint Institute for High Temperature Russian Academy of Science. E-mail: 666zhuk@ihed.ras.ru.

Conflicts of Interest: The authors declare no conflict of interest.

References

1. Ilyukhina, A.V.; Kleymenov, B.V.; Zhuk, A.Z. Development and study of aluminum-air electrochemical generator and its main components. *J. Power Sources* **2017**, *342*, 741–749. [CrossRef]
2. Sun, Y.; Liu, X.; Jiang, Y.; Li, J.; Ding, J.; Hu, W.; Zhong, C. Recent advances and challenges in divalent and multivalent metal electrodes for metal–air batteries. *J. Mater. Chem. A* **2019**, *7*, 18183–18208. [CrossRef]
3. Zhang, T.; Tao, Z.; Chen, J. Magnesium–air batteries: From principle to application. *Mater. Horiz.* **2014**, *1*, 196–206. [CrossRef]
4. Tong, F.; Wei, S.; Chen, X.; Gao, W. Magnesium alloys as anodes for neutral aqueous magnesium-air batteries. *J. Magnes. Alloys* **2021**, *9*, 1861–1883. [CrossRef]
5. Chen, X.; Liu, X.; Le, Q.; Zhang, M.; Liu, M.; Atrens, A. A comprehensive review of the development of magnesium anodes for primary batteries. *J. Mater. Chem. A* **2021**, *9*, 12367–12399. [CrossRef]
6. Wang, N.; Mu, Y.; Xiong, W.; Zhang, J.; Li, Q.; Shi, Z. Effect of crystallographic orientation on the discharge and corrosion behaviour of AP65 magnesium alloy anodes. *Corros. Sci.* **2018**, *144*, 107–126. [CrossRef]
7. Liu, X.; Xue, J.; Liu, S. Discharge and corrosion behaviors of the α-Mg and β-Li based Mg alloys for Mg-air batteries at different current densities. *Mater Des.* **2018**, *160*, 138–146. [CrossRef]
8. Ma, Y.; Li, N.; Li, D.; Zhang, M.; Huang, X. Performance of Mg–14Li–1Al–0.1 Ce as anode for Mg-air battery. *J. Power Sources* **2011**, *196*, 2346–2350. [CrossRef]

9. Deng, M.; Wang, L.; Höche, D.; Lamaka, S.V.; Jiang, P.; Snihirova, D.; Scharnagl, N.; Zheludkevich, M.L. Ca/In micro alloying as a novel strategy to simultaneously enhance power and energy density of primary Mg-air batteries from anode aspect. *J. Power Sources* **2020**, *472*, 228528–228539. [CrossRef]
10. Wang, N.; Wang, R.; Peng, C.; Peng, B.; Feng, Y.; Hu, C. Discharge behaviour of Mg-Al-Pb and Mg-Al-Pb-In alloys as anodes for Mg-air battery. *Electrochim. Acta* **2014**, *149*, 193–205. [CrossRef]
11. Feng, Y.; Lei, G.; He, Y.Q.; Wang, R.C.; Wang, X.F. Discharge performance of Mg-Al-Pb-La anode for Mg-air battery. *Trans. Nonferrous Met. Soc. China* **2018**, *28*, 2274–2286. [CrossRef]
12. Wu, Y.; Wang, Z.; Liu, Y.; Li, G.; Xie, S.; Yu, H.; Xiong, H. Improved discharge performance of Mg-6Al-7Pb alloy by microalloying with Ce. *Int. J. Electrochem. Sci.* **2018**, *13*, 10325–10338. [CrossRef]
13. Xiong, H.; Yu, K.; Yin, X.; Dai, Y.; Yan, Y.; Zhu, H. Effects of microstructure on the electrochemical discharge behavior of Mg-6wt% Al-1wt% Sn alloy as anode for Mg-air primary battery. *J. Alloys Compd.* **2017**, *708*, 652–661. [CrossRef]
14. Xiong, H.; Zhu, H.; Luo, J.; Yu, K.; Shi, C.; Fang, H.; Zhang, Y. Effects of heat treatment on the discharge behavior of Mg-6 wt.% Al-1 wt.% Sn alloy as anode for magnesium-air batteries. *J. Mater. Eng. Perform.* **2017**, *26*, 2901–2911. [CrossRef]
15. Wang, N.; Wang, R.; Feng, Y.; Xiong, W.; Zhang, J.; Deng, M. Discharge and corrosion behaviour of Mg-Li-Al-Ce-Y-Zn alloy as the anode for Mg-air battery. *Corros. Sci.* **2016**, *112*, 13–24. [CrossRef]
16. Cheng, S.; Cheng, W.; Gu, X.; Yu, H.; Wang, Z.; Wang, H.; Wang, L. Discharge properties of low-alloyed Mg–Bi–Ca alloys as anode materials for Mg–air batteries: Influence of Ca alloying. *J. Alloys Compd.* **2020**, *823*, 153779–153789. [CrossRef]
17. Gu, X.; Cheng, W.; Cheng, S.; Yu, H.; Wang, Z.; Wang, H.; Wang, L. Discharge behavior of Mg–Sn–Zn–Ag alloys with different Sn contents as anodes for Mg-air batteries. *J. Electrochem. Soc.* **2020**, *167*, 020501–020512. [CrossRef]
18. Jensen, B.W.; Gaadingwe, M.; Macfarlane, D.R.; Forsyth, M. Control of magnesium interfacial reactions in aqueous electrolytes towards a biocompatible battery. *Electrochim. Acta* **2008**, *53*, 5881–5884. [CrossRef]
19. Sathyanarayana, S.; Munichandraiah, N. A new magnesium—air cell for long-life applications. *J. Appl. Electrochem.* **1981**, *11*, 33–39. [CrossRef]
20. Deng, M.; Wang, L.; Vaghefinazari, B.; Xu, W.; Feiler, C.; Lamaka, S.V.; Hoeche, D.; Zheludkevich, M.; Snihirova, D. High-energy and durable aqueous magnesium batteries: Recent advances and perspectives. *Energy Storage Mater.* **2021**, *43*, 238–247. [CrossRef]
21. Vaghefinazari, B.; Höche, D.; Lamaka, S.V.; Snihirova, D.; Zheludkevich, M. Tailoring the Mg-air primary battery performance using strong complexing agents as electrolyte additives. *J. Power Sources* **2020**, *453*, 227880–227891. [CrossRef]
22. Wang, L.; Snihirova, D.; Deng, M.; Vaghefinazari, B.; Lamaka, S.V.; Höche, D.; Zheludkevich, M.L. Tailoring electrolyte additives for controlled Mg-Ca anode activity in aqueous Mg-air batteries. *J. Power Sources* **2020**, *460*, 228106–228118. [CrossRef]
23. Li, W.; Li, C.; Zhou, C.; Ma, H.; Chen, J. Metallic magnesium nano/mesoscale structures: Their shape-controlled preparation and Mg/air battery applications. *Angew. Chem. Int. Ed.* **2006**, *45*, 6009–6012. [CrossRef]
24. You, X.; Zhang, X.; Yu, C.; Chen, Y.; Li, H.; Hou, Y.; Tian, L.; Yang, N.; Xie, G. Corrosion protection of magnesium alloys anode by cerium-based anodization coating in magnesium-air battery. *J. Rare Earth* **2022**, *41*, 471–476. [CrossRef]
25. Hamle, R.P.; Jerabek, E.C.; Ruzzo, J.C.; Siwek, E.G. Anodes for refuelable magnesium-air batteries. *J. Electro-Chem. Soc.* **1969**, *116*, 1588–1592.
26. Available online: https://www.azom.com/article.aspx?ArticleID=6707 (accessed on 25 January 2023).
27. Zhuk, A.Z.; Sheindlin, A.E.; Kleymenov, B.V.; Shkolnikov, E.I.; Lopatin, M.Y. Use of low-cost aluminum in electric energy production. *J. Power Sources* **2006**, *157*, 641–649. [CrossRef]
28. Liu, X.; Xue, J. The role of Al2Gd cuboids in the discharge performance and electrochemical behaviors of AZ31-Gd anode for Mg-air batteries. *Energy* **2019**, *189*, 116314–116322. [CrossRef]
29. Lamaka, S.V.; Vaghefinazari, B.; Mei, D.; Petrauskas, R.P.; Höche, D.; Zheludkevich, M.L. Comprehensive screening of Mg corrosion inhibitors. *Corros. Sci.* **2017**, *128*, 224–240. [CrossRef]
30. Höche, D.; Lamaka, S.V.; Vaghefinazari, B.; Braun, T.; Petrauskas, R.P.; Fichtner, M.; Zheludkevich, M.L. Performance boost for primary magnesium cells using iron complexing agents as electrolyte additives. *Sci. Rep.* **2018**, *8*, 7578. [CrossRef]
31. Li, Q.; Xiong, W.; Yu, M.; Li, J.; Liu, L.; Zhu, G.; Wang, L.; Wang, J.; Yu, S.; Liu, E. Effect of Ce content on performance of AZ31 magnesium alloy anode in air battery. *J. Alloys Compd.* **2021**, *891*, 161914–161923. [CrossRef]

energies

MDPI

Article

One-Pot Fast Electrochemical Synthesis of Ternary Ni-Cu-Fe Particles for Improved Urea Oxidation

Marta Wala-Kapica [1,*], Aleksander Gąsior [2], Artur Maciej [2], Szymon Smykała [3], Alicja Kazek-Kęsik [2], Mehdi Baghayeri [2,4] and Wojciech Simka [2,*]

1 Faculty of Mechanics, Wroclaw University of Science and Technology, 50-370 Wroclaw, Poland
2 Faculty of Chemistry, Silesian University of Technology, 44-100 Gliwice, Poland
3 Faculty of Mechanics and Technology, Silesian University of Technology, 44-100 Gliwice, Poland
4 Department of Chemistry, Faculty of Science, Hakim Sabzevari University, Sabzevar 96131, Iran
* Correspondence: marta.wala-kapica@pwr.edu.pl (M.W.-K.); wojciech.simka@polsl.pl (W.S.)

Abstract: The climate crisis has become the most serious concern of human beings and environments worldwide in the 21st century. Global concerns about cancer epidemiology mainly originate from anthropogenic activities, particularly fossil-based operations. A key solution to this problem is the use of fuel cells—devices—capable of the direct conversion of fuel chemical energies like urea into electricity. To make their commercialization reasonable, one of the problems that needs to be solved is the development of anodic materials. The majority of investigations on urea oxidation are based on nickel, but its inadequate activity limits the efficiency of these devices. In this work, we propose and synthesize a Ni-Cu-Fe ternary electrocatalyst for urea oxidation through a fast and facile electrodeposition method. The properties of the synthesized material are examined by Scanning Electron Microscopy (SEM) conjugated with Energy Dispersive X-ray Spectroscopy (EDS), Transmission Electron Microscopy (TEM), and X-ray Diffraction (XRD). Its electrochemical properties were also examined in a 1 M KOH solution with and without 0.15 M urea. We found that the prepared powder is active in the electro-oxidation of urea, with 1.65 $V_{vs\ RHE}$ required for a current density of 10 mA cm^{-2} and a stable potential of 2.38 $V_{vs\ RHE}$ required for 3 h of polarization at 10 mA cm^{-2}.

Keywords: environmental protection; fuel cell; urea oxidation; electrodeposition; electrocatalyst

1. Introduction

The worldwide increase in demand for green and sustainable electricity takes its origin in the need for a global paradigm shift from crude-oil- and natural-gas-based operations and economies towards a circular economy based on sustainable materials and energy resources. Fuel cells are devices enabling the direct conversion of the chemical energy resulting from chemical bonds in fuel molecules to electricity via a redox reaction. As a result of such a reaction, similar to conventional gasoline or diesel engines, fuel is oxidized, but the energy loss related to the change of the heat energy to kinetic energy can be avoided. Another solution to the hazardous emissions of fossil-based energy devices is replacing them with green and sustainable energy-saving devices based on environmentally friendly substances, such as urea, simple alcohols, or ammonia, to be used as green fuel [1,2].

The commercialization of direct fuel cells would be very beneficial from an environmental point of view. In 2023, the European Union planned to ban the sale of new cars emitting CO_2 by 2035, pursuing the Net Zero emission horizon [3]. With the growing popularity and worldwide acceptance of electric vehicles, the disadvantages of using lithium-ion batteries, like how easily they catch fire, have become more and more identified to the public as well as to decision-making organizations. However, the use of fuel cells for electricity generation in combination with an electric engine could still lure the attention of enterprises, because this approach is much more beneficial than the use of batteries. Furthermore, fuel cells are lighter than batteries, which seems very beneficial from a modern engineering point of view; additionally, they can be stored in a fuel tank, similar to those used in combustion engines. Due to the fact that long stops related to battery charging can be eliminated on long journeys and due to the already existing infrastructure costs of switching from gasoline and diesel petrol to more environmentally friendly liquid alternatives, such as simple alcohols or urea, there can be a smooth transition [2,4,5].

One of the holdbacks of the commercialization of direct fuel cells, such as the Direct Urea Fuel Cell (DUFC), is the lack of proper anodic materials. The requirements the final material must meet are wide, like it must not only be active enough and cheap in production, but also maintain its activity during use [6–8].

The standard material in urea oxidation research is nickel, a metal that can create the latter of its hydroxide in alkaline environments, which later, after an application of proper potential, can be turned to its oxyhydroxide, which is an active agent for urea oxidation. Unfortunately, nickel itself does not have activity high enough to make the use of fuel cells commercially reasonable. Different approaches have been examined to increase its activity, from the use of nickel in the form of nanomaterials to doping it with different elements and the preparation of metal organic frameworks (MOFs) [9–12].

Among popular nickel dopants, iron is a very interesting element, which can be used as a dopant for nickel electrocatalysts because of its low price, related to its high abundance on Earth. Additionally, this metal is vital for living organisms, so it is non-toxic to living organisms [13–15].

The main challenge in the preparation of the industrial-scale demand for anodic materials using popular hydrothermal methods is their price and time consumption. The solution to these problems may be the preparation of catalytic materials using the electrodeposition process. This well-known technology provides a high degree of process control from the composition of the final material to the size of the prepared particles by controlling the electrical parameters of the process [16–18].

Our approach is a mix of two mentioned above: we prepared an electrocatalyst consisting of nickel doped with copper and iron using the electrodeposition process from an electrolytic bath, which has not been described previously. The physiochemical properties of the obtained powder were examined using Scanning Electron Microscopy (SEM), Transmission Electron Microscopy (TEM), and X-ray Diffraction (XRD). The electrochemical properties of the prepared materials were examined using cyclic voltammetry (CV) and chronoamperometry (CA). The obtained material shows activity towards urea oxidation, with 1.65 $V_{vs\ RHE}$ required for 10 mA cm^{-2} and a stable potential of 2.38 $V_{vs\ RHE}$ required for 3 h of polarization to 10 mA cm^{-2}.

2. Materials and Methods

2.1. Material Synthesis

All chemicals used, nickel sulphate, copper sulphate, iron sulphate, sodium citrate, ammonia, potassium hydroxide and urea, were of analytical purity and were purchased from the Avantor (Gliwice, Poland).

The samples were prepared by applying an electrodeposition process in an electrolytic bath containing nickel copper and iron salts with the concentration described in Table 1. After dissolving all the electrolytic bath ingredients, the pH was regulated using a 25% ammonia solution, which was added while stirring until a pH of 4 was reached.

Table 1. Ingredients used in preparation of the solution for the electrolytic bath.

Ingredient	Concentration
$NiSO_4$	0.375
$CuSO_4$	0.125
$FeSO_4$	0.100
Sodium citrate	0.200

The chosen electrolytic bath was prepared according to [19], but it was modified by the addition of iron sulphate and the change of the pH to acidic, which are steps that have not been described before in the literature.

Procedures for electrodeposition of catalytic powder have previously been described for similar material [20]. Briefly, the powders were electrodeposited on copper plates, which were already ground before the process using SiC grinding paper of grade 800 and 1000, followed by cleaning with deionized water and degreasing in isopropanol using ultrasound for 5 min. Prior to the electrodeposition process, the plates' surfaces were also activated by dipping them in a 40% HNO_3 solution for 60 s. Immediately after the acid treatment, the samples were rinsed with deionized water and immersed in the electrolytic bath. The electrodeposition process was conducted in an electrolyser equipped with a mechanical stirrer and nickel anodes. The copper plate on which the powder was deposited played the role of the cathode. The current density used was 20 A dm^{-2}. After the electrodeposition, the prepared powders were taken off the copper plate using ultrasound in isopropanol. After the ultrasound treatment, the isopropanol was evaporated, and the resulting powders were collected, stored, and later used for further studies.

2.2. Material Characterization

The morphology and elemental composition of the samples was examined using a Phenom ProX (Thermo Fischer Scientific, Phenom World, Eindhoven, The Netherlands) Scanning Electron Microscope (SEM) equipped with an Energy Dispersive X-ray Spectroscopy (EDS) gun. An accelerating voltage of 15 kV was used.

Observations of the morphology and nanostructure of the synthesized materials were performed by using a Titan 80–300 (Thermo Fischer Scientific, FEI, Hillsboro, OR, USA) Transmission Electron Microscope (TEM) operating at an accelerating voltage of 300 kV. The images and selected area electron diffraction (SAED) patterns were acquired in TEM mode under parallel beam illumination. The analysed powders were fist-sonicated in ethanol. After a sufficient dispersion was obtained, one droplet of suspension was deposited on a copper TEM grid covered with a lacey carbon film. Prior to examination, for the removal of any organic contamination, the samples were treated for 3 s in argon plasma using 1020 Plasma Cleaner (E.A. Fischione Instruments, Inc., Export, PA, USA).

The phase composition of the samples was determined using a Seifert 3003 TT (RICH. SEIFERT & Co. GmbH & Co. KG, Ahrensburg, Germany) powder X-ray Diffractometer with a Cu X-ray tube ($k\lambda_1$ = 1.540598 Å, $k\lambda_2$ = 1.544426 Å, $k\beta$ = 1.39225 Å), and the powder samples were analysed three times. The samples were insensitive towards X-rays, and the results are presented as an average of the scans between 5° and 80° 2θ (step size = 0.05°).

2.3. Electrochemical Characterization

To examine the electrochemical properties of the prepared powders, catalytic ink was formulated, containing 33.3 mg of powder per 1 mL of isopropanol and a water mixture with the addition of Nafion® solution. After mixing the powder with liquids, the ink was homogenized for 1 h using ultrasound until a deep, black suspension was formed. After homogenization, 5 µL of the ink was deposited on the polished surface of the glassy carbon and dried overnight. The procedure is shown in Figure 1.

Electrochemical measurements were conducted using PARSTAT 4000A (Ametek, Princeton Applied Research, Berwyn, PA, USA) and the dedicated Versa Studio software (v. 2.52.3). The prepared sample played the role of a working electrode; a glassy carbon rod was used as a counter electrode, and the Ag | AgCl electrode filled with saturated potassium chloride was used as a reference electrode.

To make the comparison between different materials easier, all potentials in this paper are given in regard to a Reversible Hydrogen Electrode (RHE). The calculations were based on Equation (1) [21]:

$$E_{RHE} = E_{Ag|AgCl(sat\ KCl)} + E^0_{Ag|AgCl(sat\ KCl)} + 0.059 \cdot pH \quad (1)$$

where $E^0_{Ag|AgCl(sat\ KCl)}$ is the standard potential of the $Ag\,|\,AgCl_{(sat\ KCl)}$ electrode (0.197 V) [21].

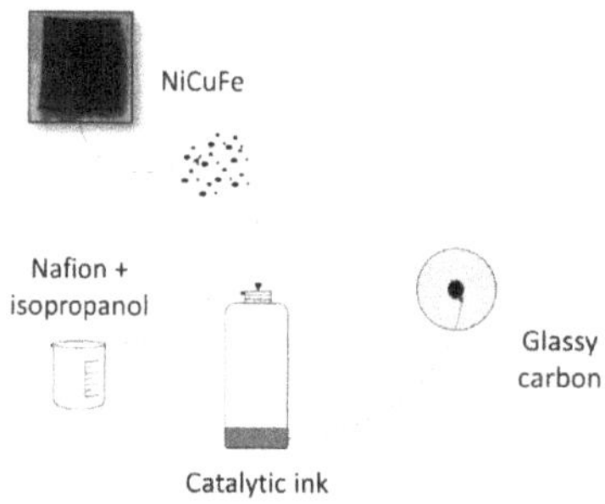

Figure 1. Schematic of the working electrode preparation used in this study.

The first step of examining the material activity was related to the selection of the potassium hydroxide concentration. Three solutions of concentrations of 0.5, 1, and 2 moles of KOH per litre were used. The activity of the prepared samples was examined using cyclic voltammetry (CV) with a scan rate of 10 mV s^{-1} in the voltage range of 0 to 0.75 V vs. $Ag\,|\,AgCl_{(sat\ KCl)}$.

The activity of the prepared powder toward urea oxidation was examined in a 1 M KOH + 0.15 M urea solution. The chosen concentration of the urea is proportional to the urea content in human urine, which is close to 0.33 M, but after adjusting to a proper pH for a Ni-based catalyst, it resulted in being diluted down.

As in the KOH solutions without the addition of urea, the activity of the prepared material was examined using cyclic voltammetry in the voltage range of 0 to 0.75 V vs. $Ag\,|\,AgCl_{(sat\ KCl)}$ with a scan rate of 10 mV s^{-1}.

An Electrochemical Impedance Spectroscopy (EIS) examination was conducted in both 1 M KOH and 0.15 M urea in 1 M KOH solutions using the same experimental setup as the rest of electrochemical measurements. Prior to the measurement, each sample was polarized to the potential of 1.32 $V_{vs\ RHE}$ for 10 min to stabilize. After that, the EIS measurement was conducted in the frequency range from 10^5 to 5×10^{-3} Hz using an amplitude of 10 mV. Registered data were fitted to the equivalent circuits using ZSimpWin software (version 3.60).

The stability of the prepared material was examined using chronoamperometry measurements (CA), where the samples were polarized for 30 min to the potential registered as the anodic peak potential during CV scans in the solution containing urea.

The stability of the proposed material was also examined by its polarization to a current density of 10 mA cm^{-2} and the recording of the resulting potential.

To examine the durability of the proposed catalyst, multiscan CV measurements were performed. Twenty subsequent scans in the voltage range from 0 to 0.75 V vs. $Ag\,|\,AgCl_{(sat\ KCl)}$ with a scan rate of 10 mV s^{-1} were conducted.

3. Results

3.1. SEM + EDX

The morphology of the prepared samples provides support for the used synthesis protocol [20,22]. It is apparent in Figure 2 that, as a result of the high current density during the electrodeposition process simultaneously with metal deposition, hydrogen gas bubbles are formed on the surface of the cathode. Actually, the bubbles that are formed on the surface of the material are playing the role of a dynamic template before they grow big enough to detach from the cathode. The entire bubble life cycle takes very little time, which makes it a perfect way to prepare electrocatalytic powders with a high active surface [22].

Figure 2. SEM image of the prepared Ni-Cu-Fe powder.

In Figure 2, an SEM image of the prepared powder is presented, which resembles a pine tree covered by snow. The obtained coral-like dendrite (Greek. *Dendron*—tree [23]) shape shows primary branches present in the centre of a particle, secondary branches radiating outwards, and ramifications to even smaller structures as the distance from the primary branch is growing. Such morphology suggests that the prepared ternary Ni-Cu-Fe material should have a high active surface, which would be a very beneficial feature considering that the catalysis is a surface process [24,25].

Figure 3 shows the EDS mapping of the prepared ternary metallic material. The results of the conducted analysis showed that all of the metals present in the electrocatalytic bath were present in the final material. The evenly distributed domains, observed in the images and approved by EDS mapping, suggest that the proposed electrolytic bath was well balanced. The most abundant element was nickel (61 $\pm$ 4%), followed by copper (22 $\pm$ 2%) and iron (10 $\pm$ 1%). The rest of the elemental composition of the material was other elements present in the electrolytic bath, such as sulphur and sodium.

3.2. XRD

In Figure 4, the XRD spectrum observed for the Ni-Cu-Fe sample is shown. Characteristic peaks related to the presence of a $Cu_{0.51}Ni_{0.49}$ alloy phase and the presence of an $FeNi_3$ alloy phase are visible. This means that during the electrodeposition process, Ni, Cu, and Fe were deposited in the form of two separate alloy phases instead of solely one or any discrete deposition of each metal.

Due to the similarity of the electrolytic bath and the process of synthesis, our materials should exhibit a similar composition as NiCu described previously [26], which is Cu2p and Ni2p, present in the form of their oxide nanoparticles with defects. The proposed material is most likely composed of nanoparticles of oxides, which formed it.

It is known that the addition of another metal to the electrocatalyst can improve its activity due to the changes in the electronic structure. The electron transfer can be promoted during the reaction [27], i.e., MoO_2 added to $NiCo_2$ improves its activity towards 5-Hydroxymethylfurfural (HMF) oxidation or leads to charge redistribution, which affects the adsorption of reaction substrates like urea or hydroxide ions [28,29]. The positive influence of copper oxide on nickel hydroxide leading to charge redistribution and thus improving the material's electrocatalytic properties was also confirmed by Density Functional Theory (DFT) [30].

Weaker Lewis acids such as Cu and Fe can shift the antibonding state of Ni toward the Fermi level, which leads to a downshift in the nickel D-band centre and thus decreases the Ni redox potential, facilitating its oxidation to NiOOH [29,31].

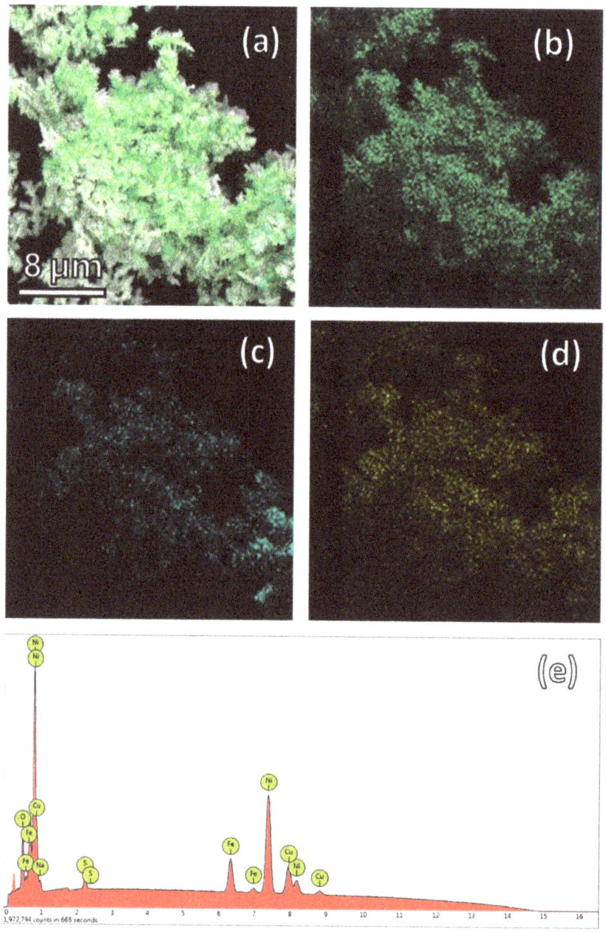

Figure 3. EDS map of the prepared material (**a**) with maps related to the presence of nickel (**b**), copper (**c**), and iron (**d**) with the EDS spectra (**e**).

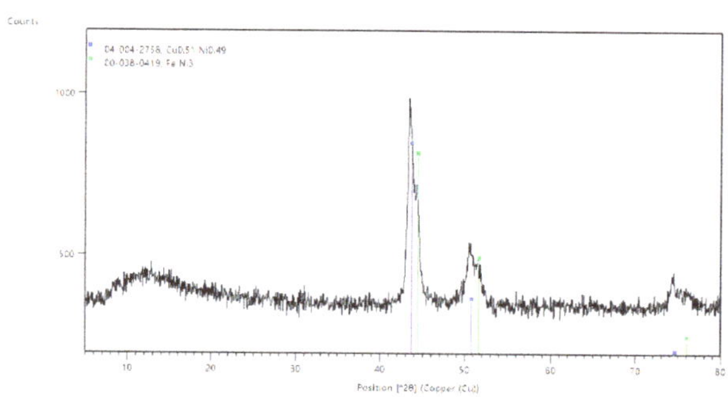

Figure 4. XRD spectra of the prepared Ni-Cu-Fe ternary powder.

3.3. TEM

Based on TEM observations, we were able to capture relevant information about the powder's morphology and structure. Even after being subjected to sonication, the obtained material consisted mainly of particles with sizes of several micrometres (Figure 5a). Observations at higher magnifications revealed a highly developed surface with dendrite-like structures (Figure 5b). The obtained SAED patterns (PDF card no. 04-004-2758 for $Cu_{0.51}Ni_{0.49}$ and card no. 00-038-0419 for $FeNi_3$) show that the powder agglomerates consist of highly polycrystalline regions (Figure 5c) as well as bulk monocrystals (Figure 5d). High-resolution imaging further confirmed the presence of both polycrystalline (Figure 5e) and bulk monocrystal (Figure 5f) areas.

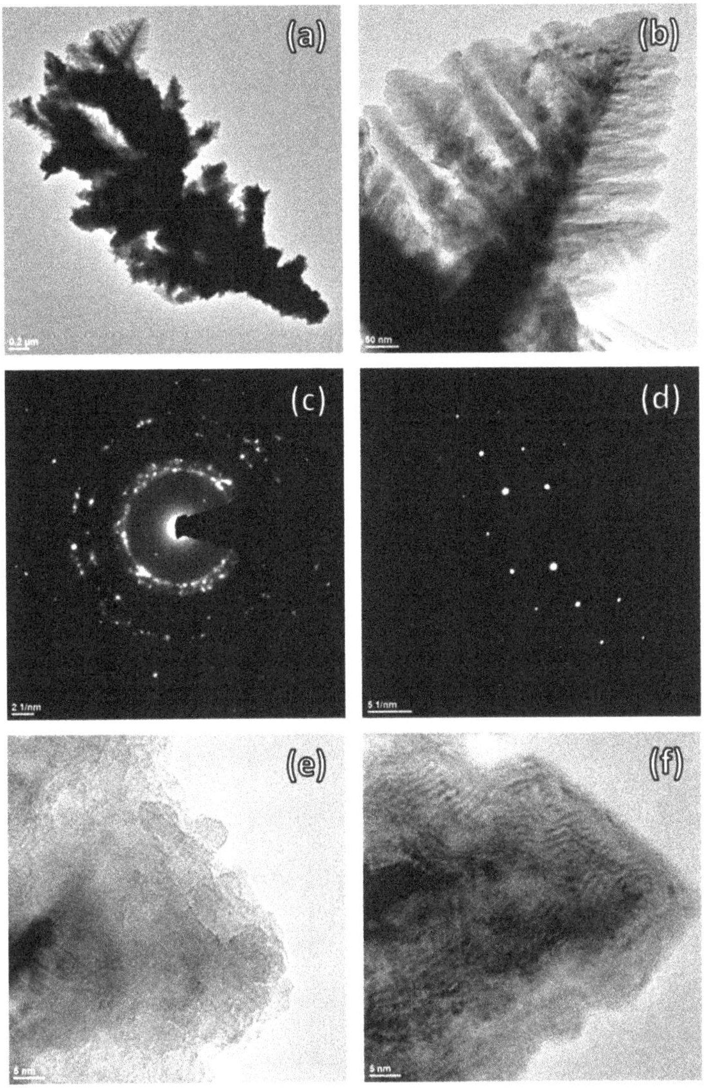

Figure 5. TEM images of Ni-Cu-Fe powder (**a**,**b**); SAED patterns of prepared sample (**c**,**d**); high-resolution TEM images of Ni-Cu-Fe powder showing polycrystalline (**e**) and monocrystal (**f**) areas.

3.4. Electrochemical Measurements

The results of cyclic voltammetry in various concentrations of potassium hydroxide are presented in Figure 6. The lowest peak current density was observed in the most diluted solution of 0.5 M concentration (see Table 2). Increasing the concentration of KOH to 1 M resulted in an increase in the anodic peak current density. Interestingly, further increase led to a decrease in the current density and a higher standard deviation caused by lower repeatability, which is why further studies were conducted in a 1 M KOH solution.

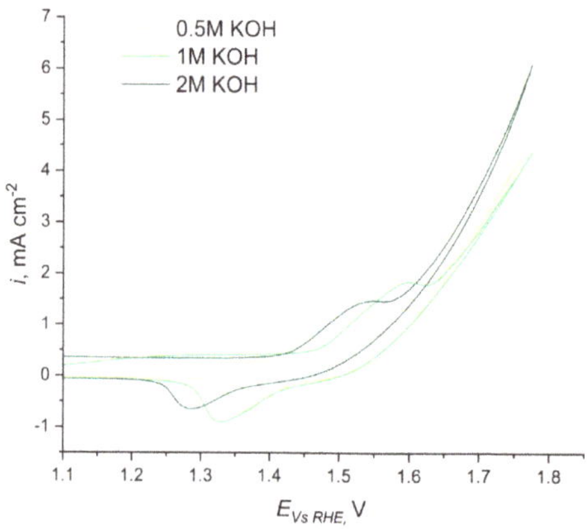

Figure 6. Comparison of the results of CV scans observed for the prepared samples in 0.5, 1, and 2 M KOH solutions.

Table 2. Comparison of the average anodic peak potentials (vs. RHE) and peak current densities observed for prepared samples in KOH solutions of various concentrations.

Solution	E_{pA}, V	i_{pA}, mA cm^{-2}
0.5 M KOH	1.560 ± 0.014	1.12 ± 0.05
1 M KOH	1.563 ± 0.029	1.42 ± 0.07
2 M KOH	1.529 ± 0.019	1.34 ± 0.17

The concentration of KOH, more precisely addressing the concentration of OH^- ions, is crucial for the activity of Ni-based catalysts due to the mechanism of their activity. After immersion into an alkaline solution, nickel hydroxide is formed, which later after polarization to its proper potential contributes to the formation of nickel oxyhydroxide, with nickel in the III valent state being the active centre for the catalysed reaction of the oxidation of simple organic molecules, such as urea.

Even more precisely speaking, during the catalysed reaction, the system needs to maintain the delicate balance between the urea and OH^- ion adsorption on the surface of the catalytic material. A too-low OH^- ion concentration causes a surface blockade with adsorbed urea without enough OH^- ions to oxidize it. If the balance is deviated in the other direction, meaning that on the electrode surface more hydroxide ions are adsorbed than urea particles, the number of exchanged electrons is lower, which results in the generation of less power by the fuel cell [32,33].

In Figure 7, a representative result of cyclic voltammetry studies of material activity toward urea oxidation is presented. The observed anodic peak current density (Table 3) is much higher than that observed in the 1 M KOH solution, which means that the prepared material is active toward urea oxidation [34].

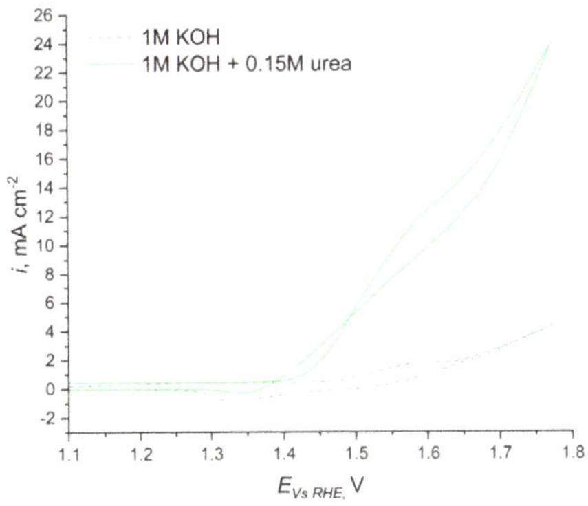

Figure 7. Comparison of the cyclic voltammograms observed for the Ni-Cu-Fe sample in the 1 M KOH solution with and without the addition of 0.15 M urea.

During the anodic scan, nickel, as well as copper and iron, are forming oxyhydroxides, which are active sites for the urea oxidation (Equations (2) and (3)) [35–37]:

$$Ni(OH)_2 + OH^- \rightarrow NiOOH + H_2O \tag{2}$$

$$NiOOH + CO(NH_2)_2 \rightarrow Ni(OH)_2 + \text{prodcuts} \tag{3}$$

The products of the UOR depend on the reaction conditions and the electrode material used [26,38].

As a result of product desorption in the voltage range related to the reaction shown in Equation (2), nickel hydroxide is being oxidized back into the oxyhydroxide and can oxidise another freshy adsorbed molecule of urea. Such a cycle of reactions can occur as long as the system is in the proper voltage range and there is enough reaction substrate in the system.

In Figure 7, two anodic peaks are observed, during both forward and reverse scans. Such a shape of a CV curve can be ascribed to the oxidation of freshly adsorbed urea during the reverse scan or to the oxidation of some intermediate chemicals formed during the forward scan, but they were not fully oxidized due to the potential swap. During the reverse scan, a cathodic peak is observed in the same potential region as in the 1 M KOH solution, but it is lower than in case of the solution without the addition of urea. This difference is probably related to the fact that nickel oxyhydroxide formed during the forward scan, being the active agent for the urea oxidation reaction (UOR); however, it then got reduced, which is why the peak related to its reduction is lower than in the case of the solution without the addition of urea.

Table 3. The observed anodic peak current density and peak potential (vs. RHE) values observed for the Ni-Cu-Fe samples compared with other materials.

Sample	E_{pA}, V	i_{pA}, mA cm^{-2}	Reference
NiCu 20	1.731	3.92	[26]
NiCuGO 20	1.736	5.88	[26]
CNT/C@FeN	1.500	3.50	[39]
$Ni_{0.9}Cu_{0.1}$	1.595	32.00	[40]
NiO/Gt 15	1.665	17.63	[41]
Ni@C	1.625	30.30	[9]
Ni-NiO/Gr-450 °C	1.525	38.24	[42]
NiCu/ZnO@MWCNTs	1.405	30.02	[43]
Ni@GO	1.475	17.10	[44]
Ni-Cu-Fe	1.695 ± 0.05	11.98 ± 0.00	In this work

GO—graphene oxide; Gt—graphite planes; Gr—graphene; MWCNT—multiwalled carbon nanotubes.

When comparing the prepared material with a similarly prepared NiCu powder catalyst and NiCuGO composite catalyst, we observed that not only was the average anodic peak potential lower, but also the peak current density registered for the Ni-Cu-Fe material was almost two times higher than for the NiCuGO composite and more than three times higher than that of NiCu material, without the addition of iron [26]. This observation suggests that the addition of iron can increase the activity of the material more efficiently with respect to the nanocarbon in the form of graphene oxide. Its presence increased the NiCu activity more efficiently than the addition of inert but well-conducting graphene oxide (GO), suggesting that the presence of iron increases the material's activity not only due to Fe conductivity, which has been observed before [45], but also due to its catalytic activity toward urea oxidation.

Electrochemical Impedance Spectroscopy comparisons between nano-sized NiFe and NiCu catalysts have shown that the urea oxidation reaction was related to lower charge transfer resistance, which means that the UOR resistance was smaller on NiFe, which explains its higher activity [34,45].

Despite that the addition of an iron NiCu electrocatalyst is still not as active as other electrocatalysts described in the literature (Table 4), a further increase in its activity can be achieved by decreasing the particle size, changing the iron content to a more beneficial Ni:Fe ratio, or adding another element, i.e., Ti [46]. P [47], F [48], S [49], or B [50].

Table 4. Comparison of the activity of different materials by comparing the potentials (vs. RHE) required for the desired current density.

Material	E@10 mA cm^{-2}, V$_{vs\ RHE}$	Urea Concentration, mol dm^{-3}	Ref.
NiCuGO20	1.720	0.15	[26]
NiCu20	1.750	0.15	
NiFeOH	1.47 for 50 mAcm^{-2}	0.33	[51]
Ni-WO$_x$	1.60 for 440 mA cm^{-2}	0.33	[52]
FeNi—oxide 2	1.380	0.33	[53]
Fe-Ni/NF	1.350	0.33	[45]
NiFe-F-4	1.44 for 50 mA cm^{-2}	0.33	[48]
Fe-NiB	1.290	0.5	[50]
Nickel carbonate	1.60 for 117.7 mA cm^{-2}	0.33	[54]
NiO$_x$/CN$_x$	1.380	0.33	[10]
Fe(EDTA)-Ni_3S_2 @NF	1.36 for 50 mA cm^{-2}	0.33	[55]
NiOOH/(LDH/αFeOOH)	1.350	0.33	[56]
$Ni_2Fe(CN)_6$	1.35 for 100 mA cm^{-2}	0.33	[57]
Ni-SO$_x$	1.65 for 323.4 mA cm^{-2}	0.33	[58]
Ir-NiFe-OH	1.38 for 100 mA cm^{-2}	0.33	[13]
NiFeTiS	1.36 for 100 mA cm^{-2}	0.33	[46]
Fe-Ni_3S_2/NF	1.3200	0.33	[45]
O$_V$-NiFe-LDH	1.300	0.33	[59]
NiFeP	1.397	0.33	[47]
NiFe	1.462	0.33	[47]
Ni-Cu-Fe	1.650 ± 0.060	0.15	In this research

The addition of iron has been proven to increase the activity of nickel electrocatalysts toward urea oxidation [45,50,53,59]. The proposed reason for this phenomenon is the stronger adsorption of urea at NiOOH active sites caused by the presence of iron, which at the same time balances the adsorption energy for intermediate products, decreasing the activation energy for NiFe urea oxidation [50]. Iron has also been proven to have a higher affinity for carbon-based urea oxidation intermediates, leading to the faster release of Ni active sites and increasing the overall reaction rate [49,50]. Another reason for the increased activity of Ni-based electrocatalysts after doping with iron is their lower charge transfer resistance [45,50,53,59].

3.5. EIS

As visible in Figure 8, in both solutions, two characteristic semicircles related to two time constants are observed. Such a shape can be attributed to the equivalent circuit, which is shown in Figure 8 and is typical for electrocatalytic reactions measured for samples prepared by drop-casting [39].

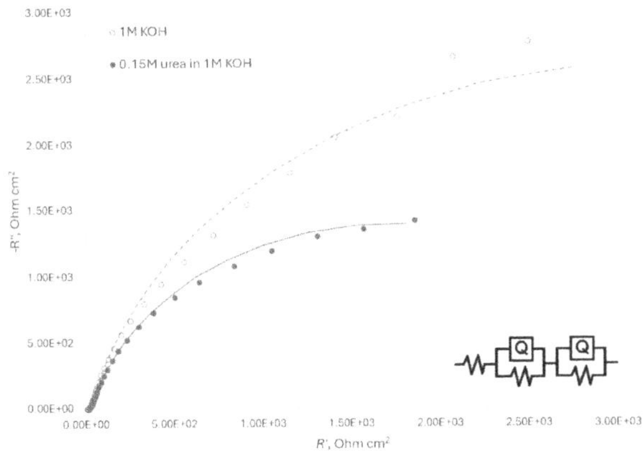

Figure 8. Nyquist plots recorded during the Electrochemical Impedance Spectroscopy investigation of the proposed material in 1 M KOH and in 1 M KOH with the addition of a 0.15 M urea solution with a graphical representation of the electrical equivalent circuit ([R(QR)(QR)]) used for the fitting.

At the beginning of the equivalent circuit, the resistance related to used aqueous solution used is measured (R_s); after that, the solution needs to diffuse through not only an electrical double layer formed after sample polarization but also through the outer layer of the electrocatalytic material deposited on the glassy carbon electrode. The final resistance is related to the charge transfer during the redox reaction (R_{CT}).

The addition of urea into the 1 M KOH solution resulted in a slight increase in the solution resistivity and an increase in the resistance related to the diffusion through the material (R_{dif}), which is caused by the bigger size of the urea particles compared to the hydroxide ions.

As expected, after the observation of higher anodic peak current densities during the CV scans, the charge transfer resistance (R_{CT}) was lower during the urea oxidation reaction (Table 5), which confirms that proposed material exhibits better electrocatalytic properties in the UOR than in water splitting [60].

Table 5. Parameters of the equivalent circuit used for EIS data fiting.

Circuit Element	1 M KOH	0.15 M Urea in 1 M KOH
R_s, Ohm cm^2	2.85 ± 0.04	3.18 ± 0.04
Q-Yo	$2.21 \pm 0.10 \times 10^{-3}$	$2.47 \pm 0.74 \times 10^{-3}$
Q-n	$8.74 \pm 5.00 \times 10^{-1}$	$7.55 \pm 1.25 \times 10^{-1}$
R_{dif}, Ohm cm^2	5743.50 ± 601.50	8549.00 ± 5191.00
Q-Yo	$9.78 \pm 2.40 \times 10^{-3}$	$5.77 \pm 5.00 \times 10^{-3}$
Q-n	$4.18 \pm 0.04 \times 10^{-1}$	$5.64 \pm 1.36 \times 10^{-1}$
R_{CT}, Ohm cm^2	31.06 ± 13.70	22.01 ± 17.44

As presented in Figure 9, during 20 subsequent CV scans in the 1 M KOH solution with the addition of 0.15 M urea, the observed peak current density values were similar to those observed during a single CV scan and remained stable during the examination, proving that the material remained stable during the experiments [48].

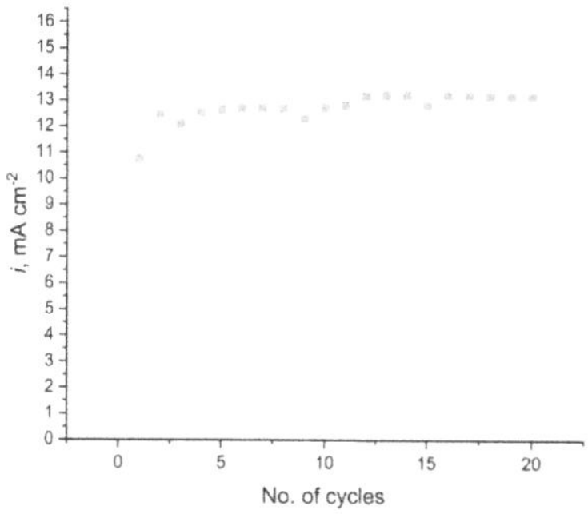

Figure 9. Comparison of the peak current densities observed for Ni-Cu-Fe samples during 20 subsequent CV scans in 1 M KOH solution with the addition of 0.15 M urea.

The material remained active in the course of chronoamperometric polarization as detected through its stable activity, with a typical CA curve for the urea electrocatalyst [20,26,34,45,53]. The observed peaks visible in Figure 10 are related to the formation of the gas bubbles of gaseous reaction products from the electrode surface [20,26,53].

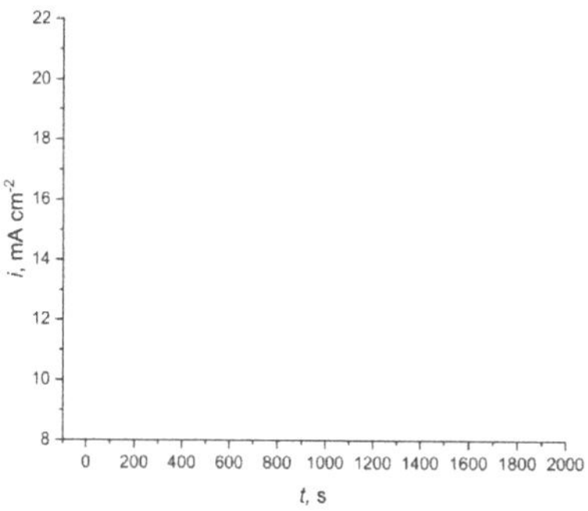

Figure 10. Results recorded during the chronoamperometric study for the sample polarized to the anodic peak potential.

After stabilization (200 s), the observed current density was similar to that observed at the anodic peaks during CV scans of 11.79 ± 1.57 mA cm^{-2}. After 600 s of polarization, the observed average current density was 12.23 ± 1.79 mA cm^{-2}, which is 104% of the initial value observed after the stabilization. At 1000 s of polarization, the observed average current density increased to 12.54 ± 0.8 mA cm^{-2}, which is 106% of the initial value. At the end of the polarization, the observed average current density was 11.88 ± 1.37 mA cm^{-2}, which is 101% of the average current density observed after the stabilization, which means that the proposed material remained stable during the examination. Such an increase in the current density observed for the material is most likely related to the oxidation of UOR intermediates adsorbed at the material surface at the beginning of polarization and as a result of the adsorption of a fresh portion of reaction substrates.

To examine the stability of the newly proposed Ni-Cu-Fe electrocatalyst, the samples were polarized to a current density of 10 mA cm^{-2}, and the resulting voltage is presented in Figure 11. In the course of polarization, which was 3 h, the observed potential varied over a narrow range of 2.36–2.37 $V_{vs\ RHE}$, which ascertained an acceptable stability for the developed electrocatalyst.

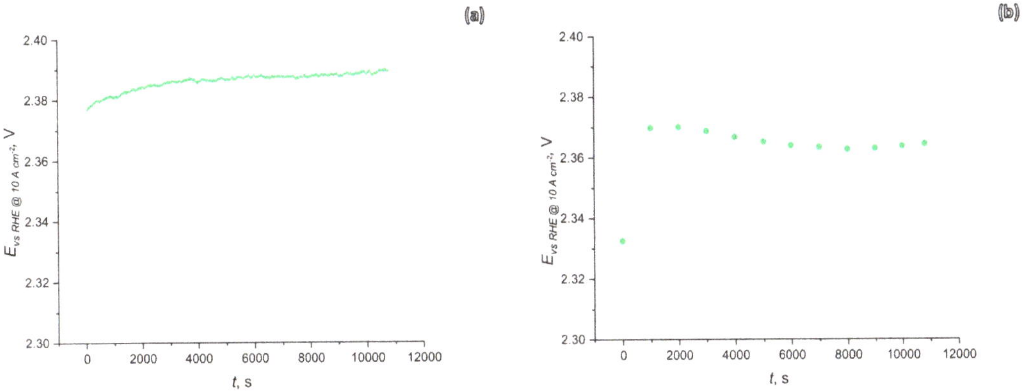

Figure 11. Comparison of the potential required for the current density of 10 mA cm^{-2}: representative example (**a**) and average values (**b**).

4. Conclusions

A simple, quick, and facile process of electrodeposition, which was kept under control, was used in the formation and collection of ternary Ni-Cu-Fe powders within an electrolytic bath with an optimized composition. As trackable evidence, all the metals presented in the electrolytic bath before the synthesis was run were detected in the obtained material, typical of the formation of a Ni-Cu-Fe complex. The prepared powder was observed under a microscope and a characteristic coral-like morphology was scanned, which poses the potential value of this ternary metallic structure for electrocatalytic applications. The obtained Ni-Cu-Fe powder was electroactive in an alkaline environment and revealed activity towards urea oxidation. To realize such a claim, the activity of the prepared material was measured, which was greater than the activity of Ni-Cu powder doped with graphene oxide. This meant that the incorporation of Fe as the tertiary metallic component into the Ni-Cu was a proper tactic for enhancing the efficiency of this hybrid structure when compared with carbon nanoforms like graphene oxide. We have to outline that the activity of the formulated material was lower than that of some of the counterparts or parent structures described in the literature; however, our ternary material was obtained through a versatile, quick, and simple method, which might be applied in developing electrocatalytic materials. However, this must be after its activity is increased upon further assessments in future research programs.

Author Contributions: Conceptualization, M.W.-K. and W.S.; methodology, M.W.-K. and W.S.; software, M.W.-K.; formal analysis, W.S.; investigation, M.W.-K., A.G., A.M., S.S. and A.K.-K.; resources, M.W.-K.; data curation, A.G., A.M., S.S. and A.K.-K.; writing—original draft preparation, M.W.-K. and S.S.; writing—review and editing, M.W.-K. and W.S.; supervision, M.B. and W.S.; project administration, W.S.; funding acquisition, M.W.-K. All authors have read and agreed to the published version of the manuscript.

Funding: This research received no external funding.

Data Availability Statement: The data presented in this study are available on request from the corresponding author. The data are not publicly available due to privacy ethical restrictions.

Conflicts of Interest: The authors declare no conflicts of interest.

References

1. Bacon, F.T. Fuel cells, past, present and future. *Electrochim. Acta* **1969**, *14*, 569–585. [CrossRef]
2. Pramuanjaroenkij, A.; Kakaç, S. The fuel cell electric vehicles: The highlight review. *Int. J. Hydrogen Energy* **2023**, *48*, 9401–9425. [CrossRef]

3. Crespo Parrondo, A.; McPhie, T. Zero emission vehicles: First 'Fit for 55' deal will end the sale of new CO_2 emitting cars in Europe by 2035. *Eur. Comm.—Press Release* **2022**, IP/22/6462. Available online: https://ec.europa.eu/commission/presscorner/detail/en/ip_22_6462 (accessed on 22 May 2024).
4. Wolf, D.D.; Smeers, Y. Comparison of Battery Electric Vehicles and Fuel Cell Vehicles. *World Electr. Veh. J.* **2023**, *14*, 262. [CrossRef]
5. Halder, P.; Babaie, M.; Salek, F.; Shah, K.; Stevanovic, S.; Bodisco, T.A.; Zare, A. Performance, emissions and economic analyses of hydrogen fuel cell vehicles. *Renew. Sustain. Energy Rev.* **2024**, *199*, 114543. [CrossRef]
6. Mansor, M.; Timmiati, S.N.; Lim, K.L.; Wong, W.Y.; Kamarudin, S.K.; Nazirah Kamarudin, N.H. Recent progress of anode catalysts and their support materials for methanol electrooxidation reaction. *Int. J. Hydrogen Energy* **2019**, *44*, 14744–14769. [CrossRef]
7. Yuda, A.; Ashok, A.; Kumar, A. A comprehensive and critical review on recent progress in anode catalyst for methanol oxidation reaction. *Catal. Rev.—Sci. Eng.* **2022**, *64*, 126–228. [CrossRef]
8. Sayed, E.T.; Eisa, T.; Mohamed, H.O.; Abdelkareem, M.A.; Allagui, A.; Alawadhi, H.; Chae, K.-J. Direct urea fuel cells: Challenges and opportunities. *J. Power Sources* **2019**, *417*, 159–175. [CrossRef]
9. Tran, T.Q.N.; Park, B.J.; Yun, W.H.; Duong, T.N.; Yoon, H.H. Metal–organic framework–derived Ni@C and NiO@C as anode catalysts for urea fuel cells. *Sci. Rep.* **2020**, *10*, 278. [CrossRef]
10. Shekhawat, A.; Samanta, R.; Panigrahy, S.; Barman, S. Electrocatalytic Oxidation of Urea and Ethanol on Two-Dimensional Amorphous Nickel Oxide Encapsulated on N-Doped Carbon Nanosheets. *ACS Appl. Energy Mater.* **2023**, *6*, 3135–3146. [CrossRef]
11. Chakrabarty, S.; Offen-Polak, I.; Burshtein, T.Y.; Farber, E.M.; Kornblum, L.; Eisenberg, D. Urea oxidation electrocatalysis on nickel hydroxide: The role of disorder. *J. Solid State Electrochem.* **2020**, *25*, 159–171. [CrossRef]
12. Li, J.; Wang, S.; Chang, J.; Feng, L. A review of Ni based powder catalyst for urea oxidation in assisting water splitting reaction. *Adv. Powder Mater.* **2022**, *1*, 100030. [CrossRef]
13. Chen, X.; Wan, J.; Chai, J.; Zhang, L.; Zhang, F.; Zhang, Q.; Gu, L.; Zheng, L.; Yu, R. Nickel-iron in the second coordination shell boost single-atomic-site iridium catalyst for hight-performance urea electrooxidation. *Nano Res.* **2024**, *17*, 3919–3926. [CrossRef]
14. Xie, J.; Qu, H.; Lei, F.; Peng, X.; Liu, W.; Gao, L.; Hao, P.; Cui, G.; Tang, B. Partially amorphous nickel-iron layered double hydroxide nanosheet arrays for robust bifunctional electrocatalysis. *J. Mater. Chem. A* **2018**, *6*, 16121–16129. [CrossRef]
15. Shit, S.; Ghosh, S.; Bolar, S.; Murmu, N.C.; Kuila, T. Understanding the Synergistic Effect in Oxygen Evolution Reaction Catalysis from Chemical Kinetics Point of View: An Iron Oxide/Nickel Oxide Case Study. *J. Electrochem. Soc.* **2020**, *167*, 116514. [CrossRef]
16. Maniam, K.K.; Chetty, R.; Thimmappa, R.; Paul, S. Progress in the Development of Electrodeposited Catalysts for Direct Liquid Fuel Cell Applications. *Appl. Sci.* **2022**, *12*, 501. [CrossRef]
17. Das, M.; Khan, Z.B.; Banerjee, M.; Biswas, A.; Dey, R.S. Three-dimensional nickel and copper-based foam-in-foam architecture as an electrode for efficient water electrolysis. *Catal. Today* **2023**, *424*, 113836. [CrossRef]
18. Wei, H.; Xu, Q.; Li, A.; Wan, T.; Huang, Y.; Cui, D.; Pan, D.; Dong, B.; Wei, R.; Naik, N.; et al. Dendritic core-shell copper-nickel alloy@metal oxide for efficient non-enzymatic glucose detection. *Sens. Actuators B Chem.* **2021**, *337*, 129687. [CrossRef]
19. Ghosh, S.K.; Grover, A.K.; Dey, G.K.; Totlani, M.K. Nanocrystalline Ni-Cu alloy plating by pulse electrolysis. *Surf. Coat. Technol.* **2000**, *126*, 48–63. [CrossRef]
20. Wala, M.M.; Szewczyk, A.K.K. Preparation of NiCuGO composite and investigation of its electrocatalytic properties in methanol oxidation. *Electrochim. Acta* **2022**, *425*, 140743. [CrossRef]
21. Niu, S.; Li, S.; Du, Y.; Han, X.; Xu, P. How to Reliably Report the Overpotential of an Electrocatalyst. *ACS Energy Lett.* **2020**, *5*, 1083–1087. [CrossRef]
22. Plowman, B.J.; Jones, L.A.; Bhargava, S.K. Building with bubbles: The formation of high surface area honeycomb-like films via hydrogen bubble templated electrodeposition. *Chem. Commun.* **2015**, *51*, 4331–4346. [CrossRef] [PubMed]
23. Kozicki, M.N. Information in electrodeposited dendrites. *Adv. Phys. X* **2021**, *6*, 1920846. [CrossRef]
24. Koh, J.H.; Won, D.H.; Eom, T.; Kim, N.K.; Jung, K.D.; Kim, H.; Hwang, Y.J.; Min, B.K. Facile CO_2 Electro-Reduction to Formate via Oxygen Bidentate Intermediate Stabilized by High-Index Planes of Bi Dendrite Catalyst. *ACS Catal.* **2017**, *7*, 5071–5077. [CrossRef]
25. Hu, B.; Yu, J.; Meng, J.; Xu, C.; Cai, J.; Zhang, B.; Liu, Y.; Yu, D.; Zhou, X.; Chen, C. Porous Ni-Cu Alloy Dendrite Anode Catalysts with High Activity and Selectivity for Direct Borohydride Fuel Cells. *ACS Appl. Mater. Interfaces* **2022**, *14*, 3910–3918. [CrossRef] [PubMed]
26. Wala, M.; Blacha-Grzechnik, A.; Stolarczyk, A.; Bajkacz, S.; Dydo, P.; Simka, W. Unexpected electrochemical oxidation of urea on a new NiCuGO composite catalyst. *Int. J. Hydrogen Energy* **2023**, *48*, 34229–34243. [CrossRef]
27. Wu, J.; Chen, J.; Yu, T.; Zhai, Z.; Zhu, Y.; Wu, X.; Yin, S. Boosting Electrochemical Kinetics of $NiCo_2$ via MoO_2 Modification for Biomass Upgrading Assisted Hydrogen Evolution. *ACS Catal.* **2023**, *13*, 13257–13266. [CrossRef]
28. Jiang, W.; Zhuo, X.; Yu, T.; Lu, J.; Zhai, Z.; Wen, H.; Yin, S. Regulating Reaction Intermediate Adsorption of $Co_{0.5}NiS_2$-Ni_3S_2 Nanorods for Efficient Urea Electrolysis. *ACS Sustain. Chem. Eng.* **2024**, *12*, 998–1006. [CrossRef]
29. Feng, B.; Jiang, W.; Deng, R.; Lu, J.; Tsiakaras, P.; Yin, S. Agglomeration inhibition engineering of nickel–cobalt alloys by a sacrificial template for efficient urea electrolysis. *J. Colloid Interface Sci.* **2024**, *663*, 1019–1027. [CrossRef]
30. Sun, H.; Li, L.; Chen, Y.; Kim, H.; Xu, X.; Guan, D.; Hu, Z.; Zhang, L.; Shao, Z.; Jung, W.C. Boosting ethanol oxidation by NiOOH-CuO nano-heterostructure for energy-saving hydrogen production and biomass upgrading. *Appl. Catal. B Environ.* **2023**, *325*, 122388. [CrossRef]
31. Taffa, D.H.; Balkenhohl, D.; Amiri, M.; Wark, M. Minireview: Ni–Fe and Ni–Co Metal–Organic Frameworks for Electrocatalytic Water-Splitting Reactions. *Small Struct.* **2023**, *4*, 2200263. [CrossRef]

32. Shi, W.; Ding, R.; Li, X.; Xu, Q.; Liu, E. Enhanced performance and electrocatalytic kinetics of Ni-Mo/graphene nanocatalysts towards alkaline urea oxidation reaction. *Electrochim. Acta* **2017**, *242*, 247–259. [CrossRef]
33. Mohamed, I.M.A.; Kanagaraj, P.; Yasin, A.S.; Iqbal, W.; Liu, C. Electrochemical impedance investigation of urea oxidation in alkaline media based on electrospun nanofibers towards the technology of direct-urea fuel cells. *J. Alloys Compd.* **2020**, *816*, 152513. [CrossRef]
34. Xu, W.; Du, D.; Lan, R.; Humphreys, J.; Wu, Z.; Tao, S. Highly active Ni-Fe double hydroxides as anode catalysts for electrooxidation of urea. *New J. Chem.* **2017**, *41*, 4190–4196. [CrossRef]
35. Gao, X.; Zhang, S.; Wang, P.; Jaroniec, M.; Zheng, Y.; Qiao, S.Z. Urea catalytic oxidation for energy and environmental applications. *Chem. Soc. Rev.* **2024**, *53*, 1552–1591. [CrossRef]
36. Hopsort, G.; Pereira Do Carmo, D.; Latapie, L.; Groenen Serrano, K.; Tzedakis, T. Progress toward a better understanding of the urea oxidation by electromediation of Ni (II)/Ni (III) system in alkaline media. *Electrochim. Acta* **2023**, *442*, 141898. [CrossRef]
37. Gnana kumar, G.; Farithkhan, A.; Manthiram, A. Direct Urea Fuel Cells: Recent Progress and Critical Challenges of Urea Oxidation Electrocatalysis. *Adv. Energy Sustain. Res.* **2020**, *1*, 2000015. [CrossRef]
38. Tatarchuk, S.W.; Medvedev, J.J.; Li, F.; Tobolovskaya, Y.; Klinkova, A. Nickel-Catalyzed Urea Electrolysis: From Nitrite and Cyanate as Major Products to Nitrogen Evolution. *Angew. Chem. Int. Ed.* **2022**, *61*, e202209839. [CrossRef]
39. Modak, A.; Mohan, R.; Rajavelu, K.; Cahan, R.; Bendikov, T.; Schechter, A. Metal−Organic polymer-derived interconnected Fe−Ni alloy by carbon nanotubes as an advanced design of urea oxidation catalysts. *ACS Appl. Mater. Interfaces* **2021**, *13*, 8461–8473. [CrossRef]
40. Hefnawy, M.A.; Fadlallah, S.A.; El-Sherif, R.M.; Medany, S.S. Synergistic effect of Cu-doped NiO for enhancing urea electrooxidation: Comparative electrochemical and DFT studies. *J. Alloys Compd.* **2021**, *896*, 162857. [CrossRef]
41. Abdel Hameed, R.M.; Medany, S.S. Enhanced electrocatalytic activity of NiO nanoparticles supported on graphite planes towards urea electro-oxidation in NaOH solution. *Int. J. Hydrogen Energy* **2017**, *42*, 24117–24130. [CrossRef]
42. Wang, S.; Xu, P.; Tian, J.; Liu, Z.; Feng, L. Phase structure tuning of graphene supported Ni-NiO Nanoparticles for enhanced urea oxidation performance. *Electrochim. Acta* **2021**, *370*, 137755. [CrossRef]
43. Basumatary, P.; Konwar, D.; Yoon, Y.S. A novel Ni–Cu/ZnO@MWCNT anode employed in urea fuel cell to attain superior performances. *Electrochim. Acta* **2018**, *261*, 78–85. [CrossRef]
44. Munde, A.V.; Mulik, B.B.; Chavan, P.P.; Sathe, B.R. Hydrothermal Syntheses of NiO−GO Nanocomposite on 3D Nickel Foam as a Support for Pt Nanoparticles and its Superior Electrocatalytic Activity Towards Methanol Oxidation. *Electrochim. Acta* **2020**, *349*, 1484–1493. [CrossRef]
45. Samdani, J.S.; Sanetuntikul, J.; Shanmugam, S. Self-supported iron-doped nickel sulfide as efficient catalyst for electrochemical urea and hydrazine oxidation reactions. *Int. J. Hydrogen Energy* **2022**, *47*, 27347–27357. [CrossRef]
46. Nur Indah Sari, F.; Ke, M.T.; Huang, Y.J.; Zheng, T.M.; Su, Y.H.; Ting, J.M. NiFe sulfide electronic structure modulation via metal doping towards enhanced urea oxidation reaction performance. *Appl. Surf. Sci.* **2024**, *649*, 159187. [CrossRef]
47. Cao, J.; Wang, G.; Hou, S.; Yu, W.; Yang, Z.; Wu, Y. Facile fabrication of amorphous NiFeP nanosheets to promote urea oxidation reaction for energy-saving hydrogen production. *J. Mater. Sci.* **2023**, *58*, 16019–16032. [CrossRef]
48. Wang, K.; Hou, M.; Huang, W.; Cao, Q.; Zhao, Y.; Sun, X.; Ding, R.; Lin, W.; Liu, E.; Gao, P. F-decoration-induced partially amorphization of nickel iron layered double hydroxides for high efficiency urea oxidation reaction. *J. Colloid Interface Sci.* **2022**, *615*, 309–317. [CrossRef]
49. Peng, K.; Liu, L.; Bhuvanendran, N.; Qiao, F.; Lei, G.; Lee, S.Y.; Xu, Q.; Su, H. Effective regulation on catalytic performance of nickel-iron-vanadium layered double hydroxide for urea oxidation via sulfur incorporation. *Mater. Adv.* **2023**, *4*, 1354–1362. [CrossRef]
50. Chen, Z.; Zheng, R.; Zou, H.; Wang, R.; Huang, C.; Dai, W.; Wei, W.; Duan, L.; Ni, B.J.; Chen, H. Amorphous iron-doped nickel boride with facilitated structural reconstruction and dual active sites for efficient urea electrooxidation. *Chem. Eng. J.* **2023**, *465*, 142684. [CrossRef]
51. Babar, P.; Patil, K.; Lee, D.M.; Karade, V.; Gour, K.; Pawar, S.; Kim, J.H. Cost-effective and efficient water and urea oxidation catalysis using nickel-iron oxyhydroxide nanosheets synthesized by an ultrafast method. *J. Colloid Interface Sci.* **2021**, *584*, 760–769. [CrossRef]
52. Wang, L.; Zhu, Y.; Wen, Y.; Li, S.; Cui, C.; Ni, F.; Liu, Y.; Lin, H.; Li, Y.; Peng, H.; et al. Regulating the Local Charge Distribution of Ni Active Sites for the Urea Oxidation Reaction. *Angew. Chem.* **2021**, *133*, 10671–10676. [CrossRef]
53. Gu, X.; Yang, D.; Liu, Z.; Wang, S.; Feng, L. Iron oxide promoted nickel/nickel oxide rough nanorods for efficient urea assisted water splitting. *Electrochim. Acta* **2020**, *353*, 136516. [CrossRef]
54. Liu, H.; Wang, P.; Qi, X.; Liu, J.; Yin, A.; Wang, Y.; Ye, Y.; Luo, J.; Ren, Z.; Yu, S.; et al. An amorphous nickel carbonate catalyst for superior urea oxidation reaction. *J. Electroanal. Chem.* **2023**, *949*, 117856. [CrossRef]
55. Lv, W.; Zhu, L.; Kong, X.; Shi, H.; Wang, C.; Zhang, R.; Wang, W. Enhancement effects of Fe-doped Ni_3S_2 on efficient electrochemical urea oxidation and mechanism insights. *J. Alloys Compd.* **2023**, *965*, 171292. [CrossRef]
56. Cai, M.; Zhu, Q.; Wang, X.; Shao, Z.; Yao, L.; Zeng, H.; Wu, X.; Chen, J.; Huang, K.; Feng, S. Formation and Stabilization of NiOOH by Introducing α-FeOOH in LDH: Composite Electrocatalyst for Oxygen Evolution and Urea Oxidation Reactions. *Adv. Mater.* **2023**, *35*, 2209338. [CrossRef]

57. Geng, S.K.; Zheng, Y.; Li, S.Q.; Su, H.; Zhao, X.; Hu, J.; Shu, H.B.; Jaroniec, M.; Chen, P.; Liu, Q.H.; et al. Nickel ferrocyanide as a high-performance urea oxidation electrocatalyst. *Nat. Energy* **2021**, *69*, 904–912. [CrossRef]
58. Gao, X.; Bai, X.; Wang, P.; Jiao, Y.; Davey, K.; Zheng, Y.; Qiao, S.Z. Boosting urea electrooxidation on oxyanion-engineered nickel sites via inhibited water oxidation. *Nat. Commun.* **2023**, *14*, 5842. [CrossRef]
59. Liu, G.; Xie, C.; Zhang, Y.; Du, Y.; Wang, J.; Lin, J.; Bai, J.; Li, J.; Zhou, C.; Zhou, T.; et al. Synergistic etching of nickel foam by Fe3+ and Cl− ions to synthesize nickel-iron-layered double hydroxide nanolayers with abundant oxygen vacancies for superior urea oxidation. *J. Colloid Interface Sci.* **2024**, *668*, 375–384. [CrossRef]
60. Golmohammadi, F.; Amiri, M. Facile synthesis of nanofiber composite based on biomass-derived material conjugated with nanoparticles of Ni–Co oxides for high-performance supercapacitors. *J. Mater. Sci. Mater. Electron.* **2020**, *31*, 2269–2279. [CrossRef]

energies

MDPI

Article

Electro-Oxidation of Glycerol on Core–Shell M@Pt/C (M = Co, Ni, Sn) Catalysts in Alkaline Medium

Rudyere Nascimento Silva [1,2,*], Leandro Aparecido Pocrifka [2], Ermete Antolini [3] and Raimundo Ribeiro Passos [2]

[1] Department of Chemistry, Environment and Food—DQA, Campus Manaus Centro, Federal Institute of Education, Science and Technology of Amazonas—IFAM, Manaus 69020-120, AM, Brazil
[2] Laboratory of Electrochemistry and Energy—LEEN, Department of Chemistry, Federal University of Amazonas—UFAM, Manaus 69088-900, AM, Brazil
[3] Scuola di Scienza dei Materiali, 16016 Cogoleto, Italy
* Correspondence: rudyere.silva@ifam.edu.br

Abstract: This study explores the development of core–shell electrocatalysts for efficient glycerol oxidation in alkaline media. Carbon-supported M@Pt/C (M = Co, Ni, Sn) catalysts with a 1:1 atomic ratio of metal (M) to platinum (Pt) were synthesized using a facile sodium borohydride reduction method. The analysis confirmed the formation of the desired core–shell structure, with Pt dominating the surface as evidenced by energy-dispersive X-ray spectroscopy (EDS). X-ray diffraction (XRD) revealed the presence of a face-centered cubic (fcc) Pt structure for Co@Pt/C and Ni@Pt/C. Interestingly, Sn@Pt/C displayed a PtSn alloy formation indicated by shifted Pt peaks and the presence of minor Sn oxide peaks. Notably, no diffraction peaks were observed for the core metals, suggesting their amorphous nature. Electrocatalytic evaluation through cyclic voltammetry (CV) revealed superior glycerol oxidation activity for Co@Pt/C compared to all other catalysts. The maximum current density followed the order Co@Pt/C > Ni@Pt/C > Sn@Pt/C > Pt/C. This highlights the effectiveness of the core–shell design in enhancing activity. Furthermore, Sn@Pt/C displayed remarkable poisoning tolerance attributed to a combined effect: a bifunctional mechanism driven by Sn oxides and an electronic effect within the PtSn alloy. These findings demonstrate the significant potential of core–shell M@Pt/C structures for designing highly active and poisoning-resistant electrocatalysts for glycerol oxidation. The presented approach paves the way for further development of optimized catalysts with enhanced performance and stability aiming at future applications in direct glycerol fuel cells.

Keywords: fuel cell application; glycerol oxidation; direct glycerol fuel cell; core–shell catalyst

1. Introduction

Fuel cells emerge as a promising alternative for obtaining clean energy, as they use renewable sources during their operation, emit practically no toxic pollutants and are highly efficient, and can be applied as sources of electrical energy for both stationary and mobile systems as well as for portable devices [1]. Among the various types of fuel cells, polymer electrolyte membrane fuel cells (PEMFCs) stand out for their versatile application potential. Basically, they operate by oxidizing a fuel at the anode (especially gaseous hydrogen) and reducing oxygen at the cathode. In some cell types, namely direct alcohol fuel cells (DAFCs), H_2 can be replaced by low molecular weight alcohols [2].

The use of glycerol as fuel in DAFCs (direct glycerol fuel cell—DGFC) has aroused the interest of researchers and has become the subject of study in recent years. Glycerol is an alcohol commonly obtained through less costly processes compared to the processes for obtaining H_2. Its low toxicity, high boiling point, high specific energy and the possibility of obtaining it from biomass are very important attractions that tend to make its application viable [3]. It is estimated that the complete oxidation of glycerol can produce 14 free electrons, which enhances its energetic effect. However, the breaking of C–C bonds poses difficulties that must be overcome before its applicability can be made viable. Therefore, several catalysts are tested to facilitate the breaking of bonds and reduce the poisoning effects frequently observed in some of the metal catalysts commonly used in the reaction [4]. Metals such as Pt, Pd, and Au (classified as precious due to their high added value) exhibit notable activity for the electro-oxidation of glycerol. However, their high cost and limited electrochemical stability pose significant challenges. Additionally, Au requires high overpotentials, making it less advantageous for use in DGFCs [5]. Bifunctional catalysts have emerged as crucial for the electro-oxidation of glycerol, as the incorporation of a second metal can enhance reaction performance and influence the mechanisms of by-product formation. Furthermore, the addition of other transition metals to precious metal catalysts can alter selectivity, enabling the targeted production of specific C3 and C2 compounds from alcohol oxidation depending on the catalyst design. Given their lower cost, catalysts composed of non-noble metals (e.g., Ni and Cu) have been investigated for the glycerol oxidation reaction (GOR). These catalysts have demonstrated high selectivity for the production of C2 and C1 compounds, particularly glycolic acid and formic acid, which are commonly obtained in the oxidation of glycerol under basic conditions using non-noble metal catalyst systems [5–7].

Falase et al. [8] investigated the glycerol oxidation on Pt-based binary and ternary alloys: the binary $Pt_{84}Ru_{16}$ and $Pt_{96}Sn_4$ catalysts showed a higher GOR activity than that of the ternary $Pt_{88}Ru_6Sn_6$ catalyst. All these catalysts demonstrated lower adsorption for the intermediates compared to the Pt/C catalyst. When studying different combinations of Pt and Sn in the form of alloys, Dou et al. [9] obtained an effective Pt_9Sn_1/C catalyst for oxidation of glycerol to glyceric acid, with the highest activity among various PtM/C (M = Sn, Mn, Fe, Co, Ni, Cu, Zn, Au) bimetallic catalysts.

When applying catalysts composed of Sn-Pt and Rh-Pt with core–shell structure (Sn@Pt and Rh@Pt, respectively) in the electro-oxidation of glycerol, Pupo et al. [10] observed an increase in catalytic performance of around 2 to 5 times higher when compared to the activity of Pt/C. Castagna et al. [11] tested different combinations of Cu@Pt in the catalysis of GOR. In that study, the researchers concluded that the combination of Pt and Cu in the core–shell format yielded catalysts with greater catalytic activity, 4 to 10 times higher when compared to the commercial catalyst PtRu/C. In addition to these, other relevant works reported core–shell catalysts with an interesting GOR performance [12–19].

On these bases, the aim of this work was to synthesize and apply nanostructured catalysts (Co@Pt/C, Ni@Pt/C, and Sn@Pt/C) for the GOR. The synthesis of these materials was carried out by deposition of Pt on previously synthesized cores. The materials were characterized by X-ray diffraction (XRD), scanning electron microscopy (SEM), energy dispersive X-ray spectroscopy (EDS), transmission electron microscopy (TEM), and cyclic voltammetry (CV). The catalytic behavior for glycerol electro-oxidation was also analyzed by CV, as well as by chronoamperometry and electrochemical impedance spectroscopy (EIS).

2. Materials and Methods

2.1. Catalyst Synthesis

Co@Pt/C, Ni@Pt/C, and Sn@Pt/C were obtained through the sequential chemical reduction of the metal precursors. Briefly, the synthesis occurred in two main stages. The first consisted of obtaining cores formed by nanoparticles of Co, Ni, and Sn. The second involved the coating of these cores with Pt followed by combination with the carbon powder support. The same synthesis sequence was applied in order to obtain catalysts with 20% metal load. The ratio between Pt and the other metals was 1:1 in each synthesis. The precursor salts of Co, Ni, or Sn ($CoCl_2 \cdot 2H_2O$–Sigma-Aldrich–Merck KGaA, Darmstadt, Germany, $NiCl_2 \cdot 6H_2O$–Sigma-Aldrich–Merck KGaA, Darmstadt, Germany, or $SnCl_2 \cdot 2H_2O$–Sigma-Aldrich–Merck KGaA, Darmstadt, Germany) were first added together with sodium citrate (Synth) in 50.0 mL of ultrapure water (18.2 MΩ cm^{-1} @ 25 °C/Milli-Q–Merck KGaA, Darmstadt, Germany). The citrate/precursor ratio was 1:1. Immediately afterwards, the mixture was heated to 90 °C under stirring while promoting a continuous flow of N_2 into the solution. After reaching the desired temperature, $NaBH_4$ (Fluka Analytical–Merck KGaA, Darmstadt, Germany) was added in excess to promote the reduction of the precursor salts. The solution was left under stirring for approximately 30 min. Then, the N_2 flow was stopped, and another 25.0 mL of ultrapure water was added to the solution containing the nanoparticles formed by Co, Ni, or Sn. Then, the appropriate amount of an aqueous solution (5%) of $H_2PtCl_6.6H_2O$ (Sigma-Aldrich) was added. Next, ascorbic acid (Chimie Test) was added in excess to promote the reduction of the Pt reducing salt, and the solution was left stirring for 2 h. Vulcan XC-72 (Cabot Corporation, Boston, MA, USA), a carbon powder with a particle size of ~30 nm and a surface area of 250 m^2 g^{-1}, was added in order to provide supported catalysts with 20% metal loading. Finally, the solution was left stirring at room temperature for 24 h. The catalysts contained in the solution underwent vacuum filtration while being washed with ultrapure water. Next, drying was carried out for 2 h at 70 °C. Finally, the samples were macerated and subjected to heat treatment at 300 °C in a N_2 atmosphere for 1 h in a tubular furnace. For comparison purposes, Pt/C (20%) was also synthesized via chemical reduction with $NaBH_4$ and applied to GOR.

2.2. Catalyst Characterization

XRD measurements were performed on a SHIMADZU XRD-7000 diffractometer (Shimadzu, Kyoto, Japan) using copper Kα radiation (1.5406 Å—potential of 40 mV and current of 30 mA). The diffraction angles 2θ were varied between 30 and 90° using a step of 0.02° at a rate of 2°/min.

SEM, EDS, and TEM techniques were used to obtain information on morphology, average diameter, particle size distribution, and composition. TEM analyses were performed using an electron microscope manufactured by JEOL—model JEM-1400Flash (JEOL LTD, Tokyo, Japan). SEM and EDS measurements were performed using a VEGA 3—TESCAN electron microscope (TESCAN, Brno, Czech Republic).

2.3. Electrochemical Measurements

The electrochemical characterization by CV and the catalytic activity measurements for GOR were obtained using a potentiostat/galvanostat from Pine Research Instrumentation Inc. (Durham, NC, US), model Wave Driver 40 DC, coupled to a microcomputer. For all measurements, KOH electrolyte (Sigma-Aldrich) with a concentration of 0.5 mol L^{-1} was used. A platinum plate electrode was used as the counter electrode. A reversible hydrogen electrode (RHE) was used as the reference electrode. A glassy carbon electrode (diameter = 5 mm) was also used as the working electrode. Before all measurements, the working electrode was polished with alumina in different particle sizes: 3.0 μm, 1.0 μm, and 0.3 μm. To deposit the catalysts on the working electrode, paints were prepared with the following proportions: 1 mg of catalyst, 200 μL of isopropyl alcohol, and 10 μL of a 5% (mass/volume) Nafion solution. This suspension remained in ultrasound for 30 min. Finally, 15 μL of this paint were applied to the surface of the glassy carbon electrode seeking its complete coverage. First, the catalysts were characterized by CV. Subsequently, CV measurements were performed in the presence of 1.0 mol L^{-1} of glycerol (Sigma-Aldrich–Merck KGaA, Darmstadt, Germany). Chronoamperometry analyses were performed by applying 0.6 V for a period of 1800 s. The EIS analyses, in turn, were performed using the potentiostat/galvanostat from the manufacturer Metrohm—model Autolab PGSTAT302N (Utrecht, The Netherlands)—with the following parameters: potential of 0.6 V, frequency range from 10 kHz to 10 mHz, 10 points per decade, and RMS of 10 mV. The results of the electrochemical measurements were normalized by the electroactive surface area (ECSA) of the catalysts, which was determined by the ratio of the charge related to the hydrogen desorption region (measured in μC) with the charge density required to reduce a monolayer of protons on a polycrystalline Pt surface (210 μC cm^{-2}).

3. Results and Discussion

3.1. Structural Characterization

The XRD patterns of the Co@Pt/C, Ni@Pt/C, and Sn@Pt/C catalysts are presented in Figure 1a and compared with the Pt/C catalyst. It can be observed that all samples presented the diffraction peaks characteristic of the face-centered cubic (fcc) structure of Pt, with 2θ values of approximately 40.0°, 46.4°, 67.8°, 81.5°, and 86.0° corresponding respectively to the planes 111, 200, 220, 311, and 222. In the XRD pattern of Sn@Pt/C, two other small peaks are observed at 2θ values of 34.1° and 51.9°, ascribed to the 101 and 211 planes of SnO_2, respectively (JCPDS No. 41-1445). For the Co@Pt/C and Ni@Pt/C samples, no other diffraction peaks are observed besides those attributed to Pt. No diffraction peaks were observed for core metals, indicating their amorphous nature. Figure 1b,c shows the diffraction peaks of the 200 and 220 planes, respectively, in which no displacement of the Pt peaks is evident, despite the presence of other metals. The 2θ values for the Co@Pt/C and Ni@Pt/C catalysts were very similar to those presented by Pt/C, indicating no alloy formation with the non-precious metals. A shift of the 200 and 220 peaks to lower angles, instead, was observed for the Sn@Pt/C catalyst, indicating the formation of a PtSn alloy. The lattice parameters of Co@Pt/C, Ni@Pt/C, Sn@Pt/C, and Pt/C catalysts are reported in Table 1. The lattice parameter of the Co@Pt/C and Ni@Pt/C was the same as Pt, indicating no alloy formation, while that of Sn@Pt/C was lower, indicating the partial formation of a PtSn solid solution. Pt peak shifts were similarly observed in the study by Cantane et al. [20], where the authors attributed this phenomenon to the partial formation of a solid solution between Pt and the secondary metal at the interface between the core and the Pt surface layer. The fcc Pt peaks in Sn@Pt/C and Pt/C catalysts were sharper than those in Co@Pt/C and Ni@Pt/C, indicating that the crystallite size of Co@Pt/C and Ni@Pt/C is larger than that of Sn@Pt/C and Pt/C. To calculate the crystallite size, the Scherrer equation was applied to the 220 peak of Pt. A proportionality constant "K" of 0.9 was used, based on the assumption of predominantly spherical particle shapes. The parameter "β" represents the full width at half maximum (FWHM) of the diffraction peak, while the wavelength "λ" was set to 0.15406 nm [11,15,21], corresponding to the Kα radiation of Cu utilized by the equipment. The crystallite size values obtained are presented in Table 1.

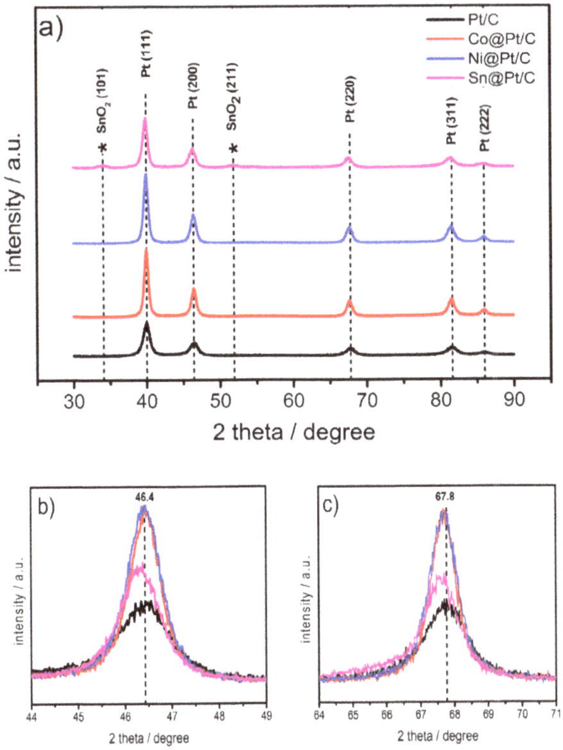

Figure 1. (**a**) XRD patterns of Co@Pt/C, Ni@Pt/C, Sn@Pt/C, and Pt/C; (**b**) comparison of the peaks of the Pt 200 fcc plane of Pt; (**c**) comparison of the peaks of the Pt 220 fcc plane. (NOTE: * identifies peaks attributed to SnO_2).

Table 1. Crystallite size (determined by XRD), lattice parameter (calculated from XRD), and particle size (measured by TEM) of the Co@Pt/C, Ni@Pt/C, Sn@Pt/C, and Pt/C catalysts.

Catalyst	XRD		TEM
	Crystallite Size (nm) [a]	Lattice Parameter (nm) [b]	Particle Size (nm)
Co@Pt/C	12.0	0.391	11.2
Ni@Pt/C	12.0	0.391	13.9
Sn@Pt/C	8.0	0.392	8.7
Pt/C	8.0	0.391	n.d. [c]

[a] Parameter determined by the Scherrer Equation: $d = (K.\lambda)/(\beta.\cos\theta)$. [b] Parameter determined by combining the equations of Bragg's Law and the equation that relates particle size and lattice parameter for a cubic crystal structure: $a_c = d(h^2 + k^2 + l^2)$ ½. [c] TEM measurements were not performed for the sample.

Figure 2a–c shows the SEM micrographs of the Co@Pt/C, Ni@Pt/C, and Sn@Pt/C catalysts. These images reveal that the catalysts exhibit predominantly globular morphological structures with a seemingly uniform and homogeneous particle size distribution. The analysis of the surface chemical composition, performed by means of EDS, shows the largely majority presence of Pt on the surface of the materials. A similar result was found for Ni@Pt by Habibi and Ghaderi [22].

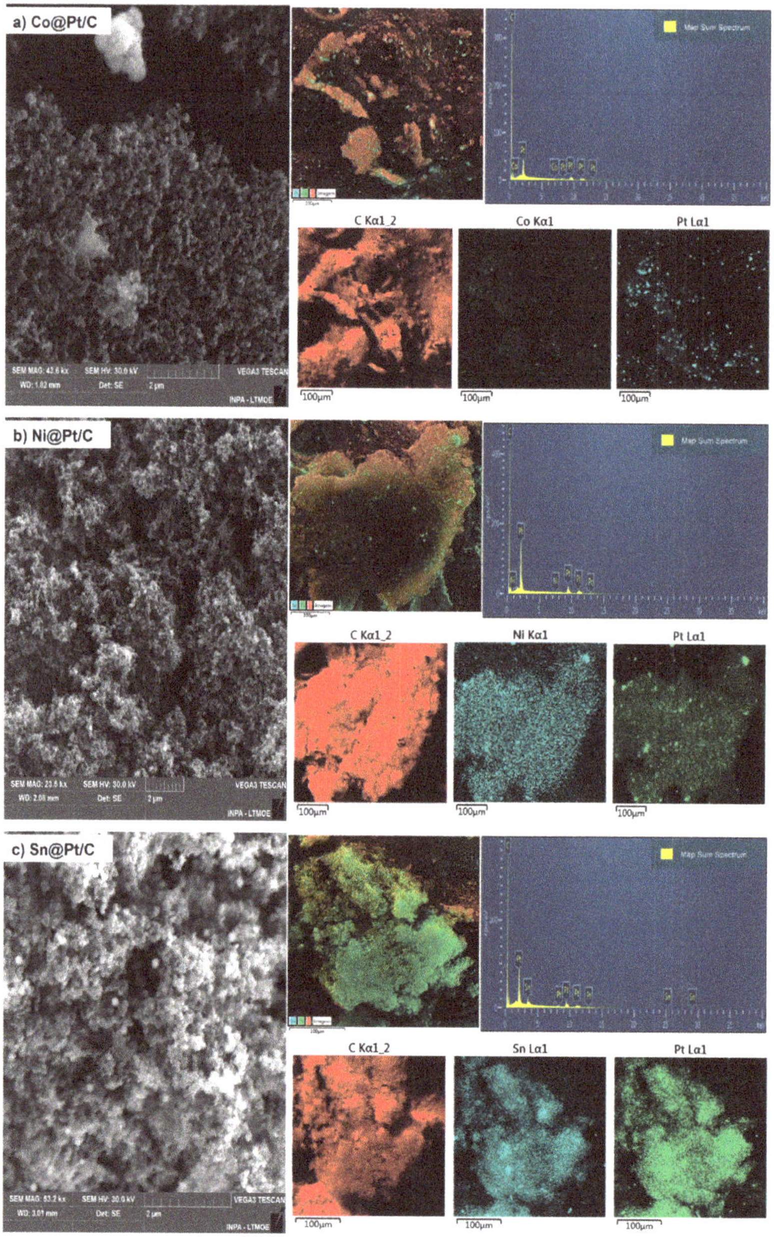

Figure 2. SEM micrographs and EDS mapping of (**a**) Co@Pt/C, (**b**) Ni@Pt/C, and (**c**) Sn@Pt/C catalysts.

Figure 3a–c presents the TEM analysis results for the Co@Pt/C, Ni@Pt/C, and Sn@Pt/C catalysts. The particles exhibit a predominantly spherical shape and satisfactory dispersion on the carbon support. However, regions with particle agglomeration were also identified. The particle size distribution histograms provide a quantitative assessment of the particle sizes, with average values listed in Table 1. A notable congruence is observed between the particle sizes determined via TEM and the crystallite sizes calculated from XRD measurements.

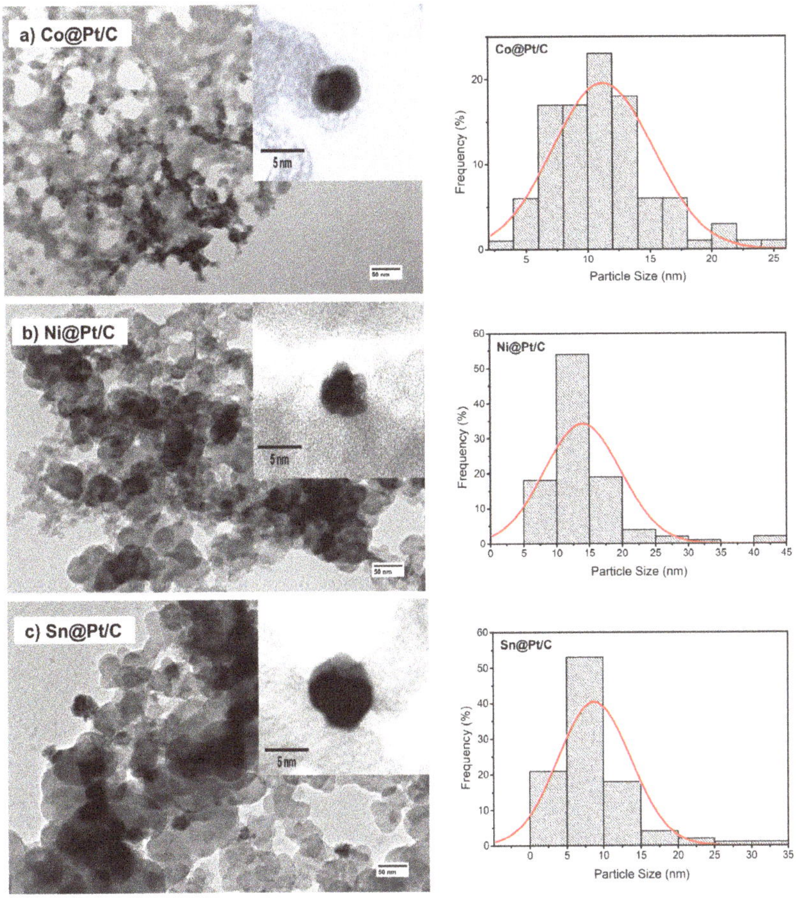

Figure 3. TEM micrographs and corresponding particle size distribution histograms for the Co@Pt/C, Ni@Pt/C, and Sn@Pt/C catalysts.

3.2. Electrochemical Characterization

The Co@Pt/C, Ni@Pt/C, and Sn@Pt/C catalysts were subjected to cyclic voltammetry analysis in 0.5 M KOH electrolyte saturated with N_2 at room temperature and compared with the Pt/C catalyst. Figure 4 shows the profiles obtained for each sample using a scan rate of 50 mV s^{-1} in a potential range from 0.05 to 1.1 V vs. RHE. The voltammograms obtained for the Co@Pt/C, Ni@Pt/C, and Sn@Pt/C catalysts showed the characteristic profiles for Pt-based materials. Three well-defined regions can be observed: (i) 0.05 to 0.4 V vs. RHE, known as the hydrogen adsorption/desorption region; (ii) 0.4 to 0.8 V vs. RHE, corresponding to non-faradaic processes, which originate from the charging of the electric double layer; and (iii) 0.8 to 1.1 V vs. RHE, referred to as the oxygen region, associated with the OH^- adsorption process. In the cathodic scan, it is possible to observe peaks around 0.75 V vs. RHE, which is related to the reduction of oxides formed in the anodic scan [23]. The absence of peak shifts in the X-ray diffraction (XRD) patterns of Co@Pt/C, Ni@Pt/C, and Sn@Pt/C compared to Pt/C suggests that the surface of these catalysts is primarily composed of Pt. This is further supported by the presence of hydrogen adsorption/desorption regions in their voltammograms, similar to those observed for polycrystalline Pt. However, the incorporation of Co, Ni, and Sn within the core likely influences the formation of the electrical double layer at the electrode surface, as evidenced by the changes observed in the voltammetric profiles [24–28].

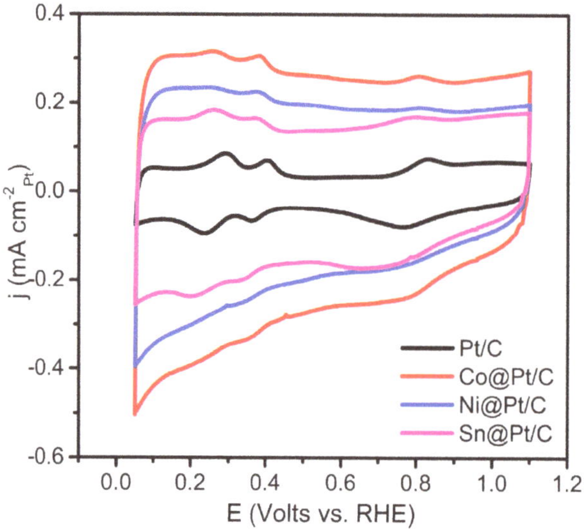

Figure 4. Cyclic voltammograms of the Co@Pt/C, Ni@Pt/C, and Sn@Pt/C catalysts. Electrolyte: 0.5M KOH saturated with N_2; scan rate: 50 mV s^{-1}.

The electrochemical activity for the glycerol oxidation of Co@Pt/C, Ni@Pt/C, Sn@Pt/C, and Pt/C catalysts was investigated by CV measurements in KOH solution + 1.0 M glycerol at room temperature, and the results are shown in Figure 5 and Table 2. As shown in Figure 5 and Table 2, the onset potential for glycerol oxidation (GOR) on Co@Pt/C is lower than that of Pt/C. This suggests that the reaction on Co@Pt/C requires less activation energy. In contrast, Ni@Pt/C and Sn@Pt/C do not exhibit a decrease in the onset potential compared to Pt/C. Interestingly, all three modified catalysts (Co@Pt/C, Ni@Pt/C, and Sn@Pt/C) show higher current density peaks in both the anodic and cathodic scans compared to Pt/C. The maximum current density follows the order Co@Pt/C > Ni@Pt/C > Sn@Pt/C > Pt/C. Since GOR produces C3, C2, and C1 products, the higher current densities observed in Figure 5 and Table 2 likely correspond to a greater degree of oxidation [3,5].

Table 2. ECSA, glycerol oxidation onset potential (E_{onset}), peak current density in the anodic scan (I_f) and peak current density in the cathodic scan (I_b) by CV; steady state current density (I_{ss}) by CA for the Co@Pt/C, Ni@Pt/C, and Sn@Pt/C catalysts.

Catalyst	ECSA ($m^2\ g^{-1}$)	E_{onset} (V)	I_f (CV) (mA cm^{-2})	I_b (CV) (mA cm^{-2})	I_{ss} (CA) (mA cm^{-2})
Co@Pt/C	3.17	0.38	3.619	2.109	0.135
Ni@Pt/C	3.47	0.44	1.993	1.042	0.107
Sn@Pt/C	6.42	0.42	1.413	0.611	0.121
Pt/C	17.89	0.42	0.732	0.539	0.106

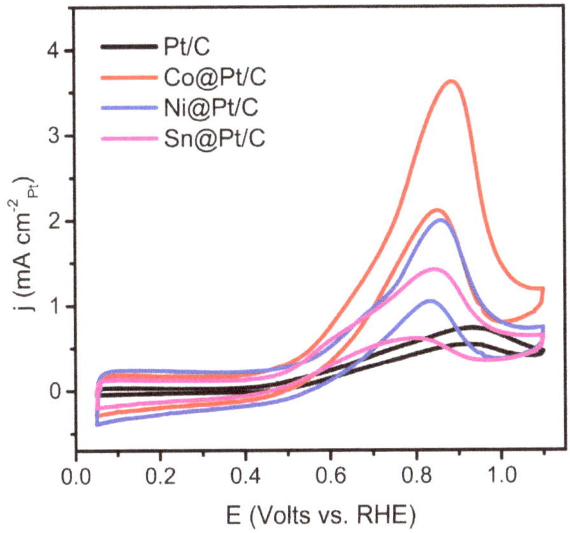

Figure 5. Voltammograms of the Co@Pt/C, Ni@Pt/C, and Sn@Pt/C catalysts compared with Pt/C. Electrolyte: 0.5 M KOH + 1.0 M glycerol. Scan rate: 50 mV s^{-1}.

The improvement in catalytic activity observed for the oxidation reactions in this study can be clearly attributed to the unique core–shell structure formed. The combination of Pt with the core metals induces electronic effects by modifying the Pt d-band energy sublevels, enhancing its catalytic properties. Additionally, geometric effects resulting from the rearrangement of the Pt crystal structure create more active catalytic sites for the glycerol oxidation reaction (GOR) [29,30].

The superior performance of the Co@Pt/C catalyst, as observed in the CV analyses with glycerol, can be attributed to its lower tendency to form Pt oxides on the surface. This behavior is evident in the CV graphs presented in Figure 4, where Co@Pt/C exhibits a delayed onset of oxide formation during the anodic scan compared to the other catalysts. Moreover, during the cathodic scan, Co@Pt/C shows earlier reduction of oxides. The ability to control the formation and reduction of oxides on the Pt surface is critical for achieving optimal performance in alcohol electrocatalysis.

The addition of Co, Ni, and Sn contributes to a reduced interaction energy between the catalytic sites and adsorbed species, further enhancing catalytic efficiency [31]. This synergistic interaction between the Pt shell and the core metals highlights the significance of the core–shell design in optimizing the activity and stability of the catalysts for GOR.

Chronoamperometric experiments were conducted at 0.6 V in a 0.5 M KOH solution containing glycerol over a period of 1800 s for the Co@Pt/C, Ni@Pt/C, Sn@Pt/C, and Pt/C catalysts. These tests aimed to evaluate the electrocatalytic activity under steady-state conditions as well as the extent of poisoning of the active catalytic surface. The current–time curves for Co@Pt/C, Ni@Pt/C, Sn@Pt/C, and Pt/C catalysts are shown in Figure 6a. In all CA curves the current drops with elapsed time; this occurs rapidly at the beginning and then becomes relatively stable. Between approximately 0 and 100 s, a sharp decline in current densities is observed for all samples. This behavior is primarily attributed to the consumption and subsequent decrease in the concentration of chemical species in the diffusion-limited layer near the electrode surface. After 100 s, a more gradual but continuous decay is observed until reaching a stable zone. This can be explained by the occupation of active sites by by-products formed during the glycerol oxidation reaction [32].

The continuous decay after stabilization at 100 s is likely due to the poisoning of surface active sites. Initially, the active sites are free from adsorbed glycerol molecules, but as the glycerol molecules adsorb again, their re-adsorption is hindered by intermediate species formed during the initial period, which leads to the poisoning of the catalytic sites. As can be seen in Figure 6b and Table 2, the order of GOR activity at the steady state was Co@Pt/C > Sn@Pt/C > Ni@Pt/C≈Pt/C.

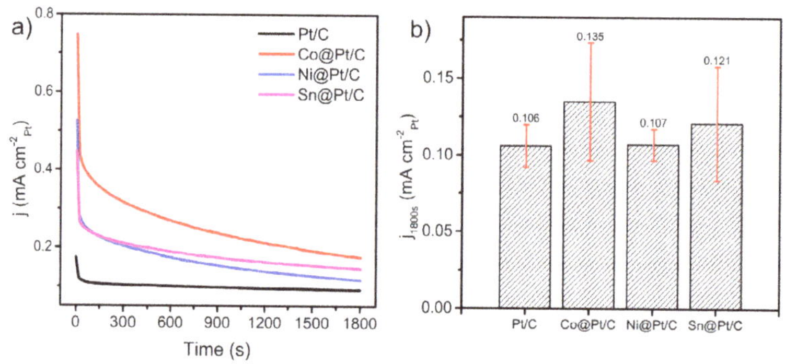

Figure 6. (**a**) Chronoamperometry at 0.6 V vs. RHE for 1800 s for the Co@Pt/C, Ni@Pt/C, Sn@Pt/C, and Pt/C catalysts. Electrolyte: 0.5 M KOH + 1.0 M glycerol; (**b**) current density values after 1800 s.

Thus, catalysts are crucial components in many chemical processes, but their performance can be hampered by the presence of intermediate species. These intermediate species can act as poisons, hindering the catalyst's ability to drive the desired reaction efficiently. To assess a catalyst's tolerance to such poisoning, two key electrochemical techniques are employed: cyclic voltammetry (CV) and chronoamperometry (CA).

CV provides valuable insights into a catalyst's electrochemical behavior. By scanning the applied potential across a set range and monitoring the resulting current response, CV offers a snapshot of the catalyst's activity. The ratio between the forward peak current (I_f) and the backward peak current (I_b), known as the I_f/I_b ratio, is a valuable parameter for evaluating the catalyst's resistance to poisoning by adsorbed intermediates. A higher I_f/I_b ratio indicates greater tolerance. This suggests that the catalyst is less susceptible to deactivation by intermediate species during the reaction cycle. On the other hand, CA takes a different approach. It measures the current response of a catalyst under a constant applied voltage over time. By analyzing the steady-state current (I_{ss}), it is possible to gain insights into the balance between glycerol adsorption and the oxidation of adsorbed intermediates. A higher I_{ss} suggests more active sites available for glycerol conversion, implying a lower degree of poisoning. The percentage of current retention at the steady state ($R = I_{ss}/I_o$, where I_o is the initial current density) is another useful parameter. A higher R value indicates that the catalyst retains a greater proportion of its initial activity over time, suggesting enhanced resistance to poisoning.

In this context, Sn@Pt/C, Ni@Pt/C, and Co@Pt/C core–shell catalysts have shown promising results. As shown in Figure 7, the Sn@Pt/C catalyst exhibits the highest I_f/I_b ratio among the three, followed by Ni@Pt/C and Co@Pt/C. This superior tolerance in Sn@Pt/C can be attributed to two factors: a bifunctional mechanism involving Sn oxides and an electronic effect between platinum and tin in the PtSn alloy [33].

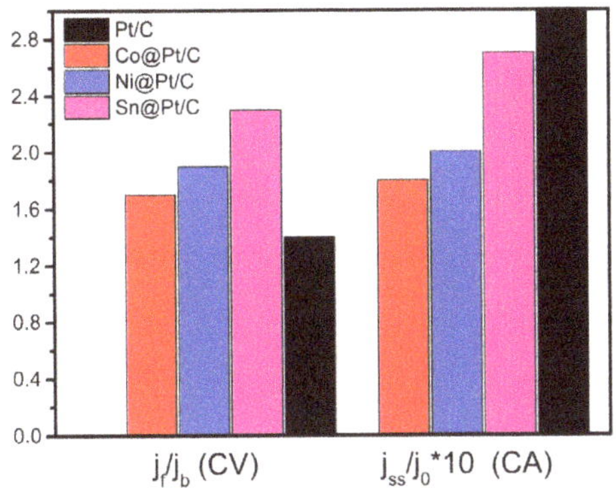

Figure 7. Histograms of I_f/I_b e R (I_{ss}/I_o) values for the Co@Pt/C, Ni@Pt/C, Sn@Pt/C, and Pt/C catalysts.

Interestingly, while the R values for all core–shell catalysts were lower than that of Pt/C (similar to observations for Pt/C and Pt-based alloys in Ref. [34]), the order of the R values mirrored the I_f/I_b ratios. This further confirms the superior poisoning tolerance of the Sn@Pt/C catalyst. Therefore, the results presented here showcase the promising performance of Sn@Pt/C as a catalyst for glycerol oxidation, demonstrating its enhanced resistance to poisoning compared to Ni@Pt/C, Co@Pt/C, and Pt/C. This work paves the way for further development of high-performance catalysts with improved stability and long-term operation.

Figure 8a presents Nyquist plots at a potential of 0.6 V, which reveal the resistance to the electrochemical reaction processes of the electrocatalysts. The charge transfer resistance (R_{ct}) at the electrode/electrolyte interface is reflected by the diameter of the small semicircle (quasi-semicircle) in the plots. A smaller diameter indicates lower R_{ct} and better electrical conductivity. The order of the quasi-semicircle diameter is Co@Pt/C < Ni@Pt/C < Sn@Pt/C < Pt/C. This aligns well with the results from CV and CA, suggesting that Co@Pt/C exhibits the lowest electron transfer resistance and the best electrical conductivity.

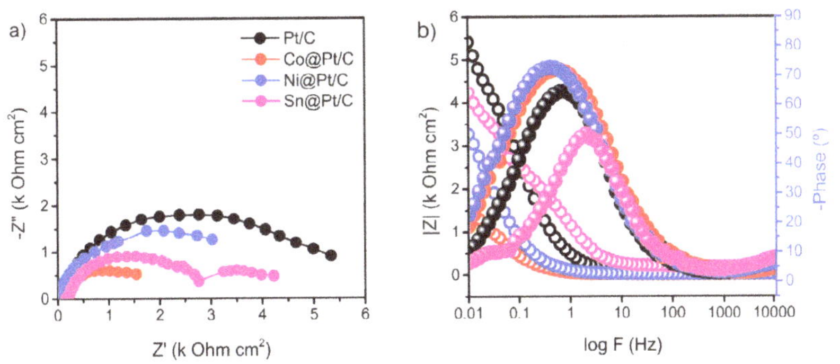

Figure 8. Nyquist diagram (**a**) and Bode diagram (**b**) of Co@Pt/C, Ni@Pt/C, Sn@Pt/C, and Pt/C catalysts. Electrolyte: 0.5 M KOH + 1.0 M glycerol. Frequency range: 10,000–0.01 Hz. Potential: 0.6 V vs. RHE.

Curiously, the Nyquist diagrams reveal additional features beyond the electrolyte resistance (R_e) and the primary charge transfer resistance (R_{ct}). Two additional charge transfer resistances ($R_{ct}1$ and $R_{ct}2$) are identified for all catalysts except Pt/C (Table 3).

Notably, the Sn@Pt/C catalyst presents a unique third semicircle, indicating a third transfer resistance value ($R_{ct}3$).

Table 3. EIS parameters for Co@Pt/C, Ni@Pt/C, Sn@Pt/C, and Pt/C. Electrolyte: 0.5 M KOH + 1.0 M glycerol. Frequency range: 10,000–0.01 Hz. Potential: 0.6 V vs. RHE.

Catalyst	R_e (Ohm)	$R_{ct}1$ (Ohm cm^2)	$R_{ct}2$ (Ohm cm^2)	$R_{ct}3$ (Ohm cm^2)
Co@Pt/C	19.2	6.5	1911.5	
Ni@Pt/C	21.5	19.9	4113.9	
Sn@Pt/C	10.4	133	2762.5	2411.5
Pt/C	23.5	5.9	5615.1	

The Bode diagram (Figure 8b) shows the charge transfer phenomena that confirm the R_{ct}s observed in Nyquist. Furthermore, it is observed that Co@Pt/C and Ni@Pt/C present a greater capacitive character, indicating a greater capacity of this catalyst to adsorb glycerol on its sites [35]. The greater capacitive character of these materials can also be observed in the voltammograms of Figure 4, in which a more prominent electric double layer region is noted in comparison to Pt/C. The Sn@Pt/C catalyst, in turn, presented greater resistance than Co@Pt/C and Ni@Pt/C, as well as a lower capacitive character. The emergence of the third R_{ct} for Sn@Pt/C can probably be associated with the SnO_2 phase, also observed in the XRD results. Thus, the phenomena observed in Bode corroborate the results presented in the initial phase of the chronoamperometry tests (Figure 6).

Overall, the EIS analysis demonstrates lower R_{ct} values for Co@Pt/C, Ni@Pt/C, and Sn@Pt/C compared to Pt/C (Table 3). This signifies faster electron transfer processes occurring on the surfaces of these modified catalysts.

4. Conclusions

This study successfully synthesized core–shell Co@Pt/C, Ni@Pt/C, and Sn@Pt/C catalysts using a $NaBH_4$ reduction method. The electrocatalytic performance of these catalysts for glycerol oxidation significantly surpassed that of Pt/C, a benchmark catalyst. Notably, Co@Pt/C exhibited the highest activity, achieving a lower onset potential and a higher maximum current density for glycerol oxidation compared to all other catalysts. Furthermore, Sn@Pt/C demonstrated superior poisoning tolerance due to a combined effect of Sn oxide-mediated bifunctional mechanism and electronic interactions within the PtSn alloy. These findings highlight the potential of core–shell structures for designing highly active and poisoning-resistant electrocatalysts for glycerol oxidation. Future work could explore optimizing the core composition and shell thickness to further enhance catalytic performance and stability.

Author Contributions: Conceptualization, R.N.S. and R.R.P.; methodology, R.N.S.; validation, L.A.P., E.A. and R.R.P.; formal analysis, R.N.S.; investigation, R.N.S. and R.R.P.; resources, L.A.P. and R.R.P.; data curation, R.N.S.; writing—original draft preparation, R.N.S.; writing—review and editing, E.A., L.A.P. and R.R.P.; visualization, E.A., L.A.P. and R.R.P.; supervision, R.R.P.; project administration, R.R.P.; funding acquisition, L.A.P. and R.R.P. All authors have read and agreed to the published version of the manuscript.

Funding: This study was financed in part by the Coordenação de Aperfeiçoamento de Pessoal de Nível Superior—Brasil (CAPES)—Finance Code 001.

Data Availability Statement: The raw data supporting the conclusions of this article will be made available by the authors upon request.

Acknowledgments: The authors would like to thank the Graduate Program in Chemistry of the Federal University of Amazonas (PPGQ-UFAM); the Central Analytical Laboratory of the Federal Institute of Education, Science and Technology of Amazonas (IFAM) for XRD analysis; the Thematic Laboratory of Optical and Electron Microscopy of the National Institute for Research in the Amazon (LMTOE/INPA) for SEM/EDS analysis; the Multiuser Center for Analysis of Biomedical Phenomena of the State University of Amazonas (CMABio/UEA) for TEM analysis; the Amazonas State Research Support Foundation (FAPEAM) for their support (Grants # 062.01163/2019-PAMEQ/2019; # 01.02.016301.03391/2021-90-EDITAL N. 010/2021—CT&I ÁREAS PRIORITÁRIAS); Coordenação de Aperfeiçoamento de Pessoal de Nível Superior—Brasil (CAPES)—Finance Code 001, and Conselho Nacional de Desenvolvimento Científico e Tecnológico—CNPq.

Conflicts of Interest: The authors declare no conflicts of interest.

References

1. Olabi, A.G.; Wilberforce, T.; Abdelkareem, M.A. Fuel cell application in the automotive industry and future perspective. *Energy* **2021**, *214*, 118955. [CrossRef]
2. Wang, Y.; Diaz, D.F.R.; Chen, K.S.; Wang, Z.; Adroher, X.C. Materials, technological status, and fundamentals of PEM fuel cells—A review. *Mat. Today* **2020**, *32*, 178–203. [CrossRef]
3. Ahmad, M.S.; Rahim, M.H.A.; Alqahtani, T.M.; Witoon, T.; Lim, J.; Cheng, C.K. A review on advances in green treatment of glycerol waste with a focus on electro-oxidation pathway. *Chemosphere* **2021**, *276*, 130128. [CrossRef]
4. Fan, L.; Liu, B.; Liu, X.; Senthilkumar, N.; Wang, G.; Wen, Z. Recent Progress in Electrocatalytic Glycerol Oxidation. *Energy Technol.* **2021**, *9*, 2000804. [CrossRef]
5. Antolini, E. Glycerol electro-oxidation in alkaline media and alkaline direct glycerol fuel cells. *Catalysts* **2019**, *9*, 980. [CrossRef]
6. Hu, X.; Lu, J.; Liu, Y.; Chen, L.; Zhang, X.; Wang, H. Sustainable catalytic oxidation of glycerol: A review. *Environ. Chem. Lett.* **2023**, *21*, 2825–2861. [CrossRef]
7. Hamada, T.; Chiku, M.; Higuchi, E.; Randall, C.A.; Inoue, H. Effect of Silver Modification on the Kinetics and Mechanisms of an Electrochemical Glycerol Oxidation Reaction at a Platinum Electrode in an Alkaline Medium. *ACS Appl. Energy Mater.* **2024**, *7*, 1970–1982. [CrossRef]
8. Falase, A.; Main, M.; Garcia, K.; Serov, A.; Lau, C.; Atanassov, P. Electrooxidation of ethylene glycol and glycerol by platinum-based binary and ternary nano-structured catalysts. *Electrochim. Acta* **2012**, *66*, 295–301. [CrossRef]
9. Dou, J.; Zhang, B.; Liu, H.; Hong, J.; Yin, S.; Huang, Y.; Xu, R. Carbon supported Pt_9Sn_1 nanoparticles as an efficient nanocatalyst for glycerol oxidation. *Appl. Catal. B Environ.* **2016**, *180*, 78–85. [CrossRef]
10. Pupo, M.M.S.; López-Suárez, F.E.; Bueno-López, A.; Meneses, C.T.; Eguiluz, K.I.B.; Salazar-Banda, G.R. Sn@Pt and Rh@Pt core–shell nanoparticles synthesis for glycerol oxidation. *J. Appl. Electroch.* **2015**, *45*, 139–150. [CrossRef]
11. Castagna, R.M.; Sieben, J.M.; Alvarez, A.E.; Duarte, M.M.E. Electrooxidation of ethanol and glycerol on carbon supported PtCu nanoparticles. *Int. J. Hydrogen Energy* **2019**, *44*, 5970–5982. [CrossRef]
12. Song, J.; Yu, J.; Zhang, M.; Liang, Y.; Xu, C. Glycerol electrooxidation on Au/Ni core/shell three-dimensional structure catalyst. *Int. J. Electrochem. Sci.* **2012**, *7*, 4362–4368. [CrossRef]
13. Fashedemi, O.O.; Miller, H.A.; Marchionni, A.; Vizza, F.; Ozoemena, K.I. Electro-oxidation of ethylene glycol and glycerol at palladium-decorated FeCo@Fe core–shell nanocatalysts for alkaline direct alcohol fuel cells: Functionalized MWCNT supports and impact on product selectivity. *J. Mater. Chem. A* **2015**, *3*, 7145–7156. [CrossRef]
14. Miller, H.A.; Bellini, M.; Vizza, F.; Hasenöhrl, C.; Tilley, R.D. Carbon supported Au–Pd core–shell nanoparticles for hydrogen production by alcohol electroreforming. *Catal. Sci. Technol.* **2016**, *6*, 6870–6878. [CrossRef]
15. Silva, L.S.R.; López-Suárez, F.E.; Perez-Cadenas, M.; Santos, S.F.; Costa, L.P.; Eguiluz, K.I.B.; Salazar-Banda, G.R. Synthesis and characterization of highly active Pb_x@Pt_y/C core-shell nanoparticles toward glycerol electrooxidation. *Appl. Catal. B Environ.* **2016**, *198*, 38–48. [CrossRef]
16. Zhou, Y.; Shen, Y. Selective electro-oxidation of glycerol over Pd and Pt@Pd nanocubes. *Electrochem. Commun.* **2018**, *90*, 106–110. [CrossRef]
17. Yang, L.; He, T.; Lai, C.; Chen, P.; Hou, Z. Selective oxidation of glycerol with oxygen in base-free solution over N-doped-carbon-supported Sb@$PtSb_2$ hybrid. *Chin. J. Catal.* **2020**, *41*, 494–502. [CrossRef]
18. Zhou, Y.; Shen, Y.; Luo, X. Optimizing the activity and selectivity of glycerol oxidation over core-shell electrocatalysts. *J. Catal.* **2020**, *381*, 130–138. [CrossRef]
19. Pu, H.; Dong, K.; Zhang, T.; Dai, H.; Wang, Y.; Deng, Y. Regulation of the shell thickness and shell components in PtCu/PdCu core–shell tripods for ethylene glycol and glycerol oxidation reactions. *J. Mater. Chem. A* **2022**, *10*, 10614–10624. [CrossRef]

20. Cantane, D.A.; Oliveira, F.E.R.; Santos, S.F.; Lima, F.H.B. Synthesis of Pt-based hollow nanoparticles using carbon-supported Co@Pt and Ni@Pt core–shell structures as templates: Electrocatalytic activity for the oxygen reduction reaction. *Appl. Catal. B Environ.* **2013**, *136–137*, 351–360. [CrossRef]
21. Oh, L.S.; Han, J.; Lim, E.; Kim, W.B.; Kim, H.J. PtCu Nanoparticle Catalyst for Electrocatalytic Glycerol Oxidation: How Does the PtCu Affect to Glycerol Oxidation Reaction Performance by Changing pH Conditions? *Catalysts* **2023**, *13*, 892. [CrossRef]
22. Habibi, B.; Ghaderi, S. Ni@Pt core-shell nanoparticles as an improved electrocatalyst for ethanol electrooxidation in alkaline media. *Iran. J. Hydrog. Fuel Cell* **2015**, *2*, 109–119. [CrossRef]
23. Sharma, R.; Gyergyek, S.; Andersen, S.M. Critical thinking on baseline corrections for electrochemical surface area (ECSA) determination of Pt/C through H-adsorption/H-desorption regions of a cyclic voltammogram. *Appl. Catal. B Environ.* **2022**, *311*, 121351. [CrossRef]
24. Silva, J.C.M.; De Souza, R.F.B.; Parreira, L.S.; Teixeira Neto, E.; Calegaro, M.L.; Santos, M.C. Ethanol oxidation reactions using SnO_2@Pt/C as an electrocatalyst. *Appl. Catal. B Environ.* **2010**, *99*, 265–271. [CrossRef]
25. Godínez-Salomón, F.; Hallen-López, M.; Solorza-Feria, O. Enhanced electroactivity for the oxygen reduction on Ni@Pt core-shell nanocatalysts. *Int. J. Hydrogen Energy* **2012**, *37*, 14902–14910. [CrossRef]
26. Reyes-Rodríguez, J.L.; Godínez-Salomón, F.; Leyva, M.A.; Solorza-Feria, O. RRDE study on Co@Pt/C coreeshell nanocatalysts for the oxygen reduction reaction. *Int. J. Hydrogen Energy* **2013**, *38*, 12634–12639. [CrossRef]
27. Habibi, B.; Ghaderi, S. Ethanol electrooxidation on the Co@Pt core-shell nanoparticles modified carbon-ceramic electrode in acidic and alkaline media. *Iran. J. Hydrog. Fuel Cell* **2016**, *1*, 19–31. [CrossRef]
28. Habibi, B.; Ghaderi, S. A comparative study of electrocatalytic performance of the M@Pt (M = Fe_3O_4, Co and Ni) nanoparticles for direct ethanol fuel cells. *J. Iran. Chem. Soc.* **2017**, *14*, 1633–1642. [CrossRef]
29. Liu, C.; Yang, F.; Schechter, A.; Feng, L. Recent progress of Ni-based catalysts for methanol electrooxidation reaction in alkaline media. *Adv. Sens. Energy Mater.* **2023**, *2*, 100055. [CrossRef]
30. Lu, Q.; Zhao, X.; Luque, R.; Eid, K. Structure-activity relationship of tri-metallic Pt-based nanocatalysts for methanol oxidation reaction. *Coord. Chem. Rev.* **2023**, *493*, 215280. [CrossRef]
31. Sieben, J.M.; Alvarez, A.E.; Sanchez, M.D. Glycerol electrooxidation on carbon-supported Pt-CuO and PtCu-CuO catalysts. *Electroc. Acta* **2023**, *439*, 141672. [CrossRef]
32. Rezaei, B.; Saeidi-Boroujeni, S.; Havakeshian, E.; Ensafi, A.A. Highly efficient electrocatalytic oxidation of glycerol by Pt-Pd/Cu trimetallic nanostructure electrocatalyst supported on nanoporous stainless steel electrode using galvanic replacement. *Electroc. Acta* **2016**, *203*, 41–50. [CrossRef]
33. Antolini, E.; Gonzalez, E.R. Effect of synthesis method and structural characteristics of Pt–Sn fuel cell catalysts on the electro-oxidation of CH_3OH and CH_3CH_2OH in acid medium. *Catal. Today* **2011**, *160*, 28–38. [CrossRef]
34. Jacob, J.M.; Corradini, P.G.; Antolini, E.; Santos, N.A.; Perez, J. Electro-oxidation of ethanol on ternary Pt–Sn–Ce/C catalysts. *Appl. Catal. B Environ.* **2015**, *165*, 175–184. [CrossRef]
35. Maya-Cornejo, J.; Diaz-Realb, J.A.; Lopez-Miranda, J.L.; Álvarez-Contreras, L.; Esparza, R.; Arjonac, N.; Estéveza, M. Formation of Cu@Pd core@shell nanocatalysts with high activity for ethanol electro-oxidation in alkaline medium. *Appl. Surf. Sci.* **2021**, *538*, 148119. [CrossRef]

energies

MDPI

Article

Prussian Blue Analogues Based on 3d-Metals as Cathode Materials for Magnesium Ion Batteries

Andrey Arbenin, Semyon Egorov, Igor Prikhodko, Anna Fedorova, Anastasia Penkova and Artem Selyutin *

Saint Petersburg State University, Universitetsky pr.26, Saint Petersburg 198504, Russia; a.penkova@spbu.ru (A.P.)
* Correspondence: a.selyutin@spbu.ru

Abstract: Prussian blue PB analogues $(K;Mg)_xM[Fe(CN)_6]$ were obtained via the co-precipitation method. Manganese, iron, cobalt, nickel, copper and zinc were selected to obtain the metal hexacyanoferrates (MHCFs) of these metals and to systematically study their structural and electrochemical properties as cathode materials. The obtained substances were characterized via X-ray powder analysis, scanning electron microscopy, thermogravimetric analysis and elemental analysis. An electrochemical study of the obtained cathode materials relative to a platinum anode in an aqueous medium and a magnesium anode in a nonaqueous medium was carried out, and the cycling parameters were determined. The influence of 3d-metal nature on the composition–structure–properties of hexacyanoferrates was demonstrated. MHCFs are promising cathode materials for Mg^{2+} intercalation/deintercalation in aqueous electrolytes.

Keywords: aqueous electrolytes batteries; intercalation; hexacyanoferrates; cathode materials

1. Introduction

Analogues of Prussian Blue with various transition metal atoms have found applications in areas such as hydrogen storage [1,2], electrochromic materials [3], wastewater treatment [4], photomagnetic materials [5] and biomedicine [6]. These compounds can also act as catalysts for organic synthesis [7,8] and as precursors for the synthesis of nanoscale oxides [9,10]. In the last 10 years, the direction of using transition metal ferrocyanides as cathode materials for water batteries has been actively developed [11]. The cubic structure of Prussian Blue analogues, with a rigid framework of transition metal atoms and a network of open channels for diffusion, promotes the reversible intercalation/deintercalation of both monovalent [12] and multivalent ions [13]. Prussian Blue analogues have a double perovskite structure with the general formula $A_xB_yM_z[Fe(CN)_6]mH_2O$ (A: alkali or alkaline earth atoms; B: vacancies; M: transition metal), where cyanide anions form a perovskite-like structure [14]. The presence of coordinated water molecules and the incorporation of various cations can lead to structural distortions and a reduction in the symmetry of the primary cubic structure of Prussian Blue compounds to a monoclinic, rhombohedral or tetragonal structure. Materials with a rhombohedral structure and a relatively low coordinated water content have a great advantage as cathodes for rechargeable batteries [14]. The low degree of deformation of the crystal lattice during ion exchange and intercalation is one of the advantages of these compounds. Other advantages of these compounds include high theoretical capacity, low cost, scalability and environmentally friendly synthesis [15]. In addition, by introducing transition metal atoms into the ferrocyanides structure that are capable of undergoing a redox reaction in an aqueous solution, it is possible to further increase the capacity of the battery. [16]. Most transition metal ferrocyanides reported in

 https://doi.org/10.3390/en18030711

the literature have micron-sized particles, which is not the optimal morphology for ion diffusion [17]. Today, lithium-ion batteries are widely used as energy storage devices. They have high power, sufficient cycling efficiency and stable operation [18]. On the other hand, the use of lithium-ion technologies for large-scale applications is very difficult, mainly due to the low stability of lithium in the atmosphere. In addition, lithium is characterized by high cost and a small and uneven distribution of reserves in the Earth's crust [19]. In its metallic state, magnesium is not as actively oxidized by atmospheric oxygen and water as lithium is. This means that magnesium-ion batteries can be operated using aqueous electrolytes, which are environmentally friendly and safer. [20]. At the same time, magnesium has similar properties to lithium due to the proximity of the ionic radii [21,22]. On the other hand, the charge of magnesium is twice as high, which gives its advantage in volume capacity [23]. Despite all these positive characteristics, the development of magnesium-ion technology is lagging behind that of lithium. Due to the higher charge of the magnesium ion (+2), a slowing down of the diffusion process of this ion through the cathode materials should be observed. As a consequence, the battery performance and the cyclability of the intercalation/deintercalation process should be reduced. Recent work shows that the low mobility of magnesium ions is due to both strong ionic interactions and the redistribution of divalently charged cations in the electrode material [24]. Several strategies have been proposed to overcome this problem:

1. Use compounds with high divalent ion mobility as cathode materials [25,26];
2. Reduce the diffusion length of magnesium ions by using nanoscale and mesoporous materials [27].

Transition metal ferrocyanides have already been well established as cathode materials for lithium- and sodium-ion batteries [28,29]. Publishing on these systems has continued to grow. Due to structural features, Prussian Blue analogues can intercalate/deintercalate multivalent ions such as Mg^{2+}, Ca^{2+}, Sr^{2+}, Ba^{2+}, Zn^{2+}, Al^{3+}, indicating a sufficiently high mobility of multivalent ions in the electrode material [30–35]. As the development of technologies based on multivalent ions is a relatively recent direction, there are few articles devoted to this topic. There are only three papers in the literature on the study of the behaviour of transition metal ferrocyanides in aqueous solutions of magnesium salts, and the data are rather descriptive [36–38]. Therefore, there is a need for fundamental research on such systems. The popularity of transition metal ferrocyanides as cathode materials for water batteries is due to a number of reasons. The properties of the cathode material based on these compounds can be adjusted by changing the composition, morphology and particle size of the substance. Without local or global changes in their crystal structure, these compounds are also capable of intercalating ions of various metals [39]. The synthesis of these compounds is inexpensive, and the materials themselves are safe to use and non-toxic [40].

In this paper we present our systematic study on the influence of various 3d-metals' nature on the electrochemical properties of PB analogues as cathode materials for magnesium-ion batteries. They have high ionic conductivity that justifies their limited voltage in aqueous electrolytes.

2. Materials and Methods

2.1. Synthesis

Nitrates of divalent metals (manganese, iron, cobalt, nickel, copper and zinc) and potassium ferrocyanide (III) produced by Vekton (Russia) were used to synthesize analogues of Prussian Blue. All chemicals were analytical grade and used without further purification. The preparation of MHCFs (M = Mn, Fe, Co, Ni, Cu and Zn) was performed using a solution precipitation method. Typically, 40 mL of 20 mM $K_3Fe(CN)_6$ in an aqueous solution and 40 mL of 40 mM $M(NO_3)_2 \cdot 6H_2O$ (M = Mn, Fe, Co, Ni, Cu and Zn) in an aqueous solution were slowly added simultaneously to 20 mL of deionized water at a same rate of 1 mL·min^{-1} under continuous stirring at 343 K. The resulting solution was kept at room temperature for 18 h. The resulting precipitate was dried in an oven at 333 K for 6 h.

After precipitation, three MHCFs (M = Mn, Ni and Zn) were filtered and washed many times with deionized water. The other MHCFs (M = Fe, Co and Cu) were centrifuged at 13,000 rpm for 3 min. This was necessary to force their precipitation. The resulting precipitate was washed five times with deionized water and centrifuged.

2.2. Materials' Characterization

The crystalline structure of the sample was characterized via the powder X-ray diffraction method (XRD) on a Bruker D2 Phaser diffractometer with Cu Kα radiation over the 2Θrange of 5−80°. Theoretical models were constructed to refine the structure of compounds using the Rietveld's method in the programme «Topas 4.2». The size and morphological appearance of the powder were observed using scanning electron microscopy (SEM) on the Hitachi S-3400N in a secondary electron registration mode with an accelerating voltage of 20 kV. Thermal material testing was performed on SETSYS Evolution 16 in the temperature range from room temperature to 700 K. The chemical composition of the PB analogues was determined the inductively coupled plasma (ICP) method in a solution of nitric acid. The measurements were performed on the Shimadzu ICPE-9000.

2.3. Electrochemical Measurements

The electrochemical intercalation and deintercalation of Mg ions were performed using a three-electrode cell. The working electrodes were fabricated by mixing 80 wt.% MHCFs, 7 wt.% polyvinylidene fluoride (Kynar) and 13 wt.% graphite and then grinding the components together with a mortar and pestle. The powder was mixed with 1-methyl-2-pyrrolidinone (PVDF) to create a slurry that was deposited onto steel or platinum foils. These electrodes were dried under vacuum at 353 K and exhibited a typical mass loading of approximately 8 mg/cm^2. AgCl/Ag was used as the reference electrode, and the counter electrode was steel or platinum. The distance between the electrodes was the same in each experiment. It was 3 cm, and the square electrodes were 1 cm wide. For the electrolyte, an aqueous solution of 1 M $MgSO_4$ was used. Cyclic voltametric measurements were carried out at a scanning rate of 1 mV/s. The working range of the potential was from 2.35 to 3.55 V vs. Mg^{2+}/Mg. The electrochemical behaviour of MHCFs was tested in galvanostatic charge/discharge mode using a potentiostat–galvanostat P-30SM Elins. In a non-aqueous solution, the discharge capacity of the synthesized cathode materials was also determined. Anhydrous acetonitrile was used as the solvent. A salt of magnesium and p-toluenesulfonic acid was chosen as the electrolyte. A two-contact cell with metallic magnesium as the anode was used for the electrochemical studies.

3. Results and Discussion

3.1. X-Ray Diffraction

Full-profile powder X-ray diffraction analyses of the samples were performed on a Bruker D2 phaser diffractometer (Cu Kα), with the precise identification of cell parameters via the Rietveld method. MHCFs were found to have a different crystal structure, although they were obtained under identical synthesis conditions.

The cubic MnHCF, FeHCF, CoHCF, NiHCF and CuHCF have a space group of Fm-3m (Figure 1). PB analogues with copper and manganese were well crystallized. The broadening of peaks in the X-ray diffraction of NiHCF, FeHCF and CoHCF may be due to a decrease in particle size or the amorphization of the structure. A small amount of the amorphized phase is present in FeHCF and CoHCF, as evidenced by the appearance of a halo at small diffraction angles. The trigonal ZnHCF has a space group of R-3c. The resulting sample is well crystallized and has no crystalline impurities (Tables 1 and 2). Figure 1 shows the experimental results for a full-profile powder X-ray diffraction analysis of the samples.

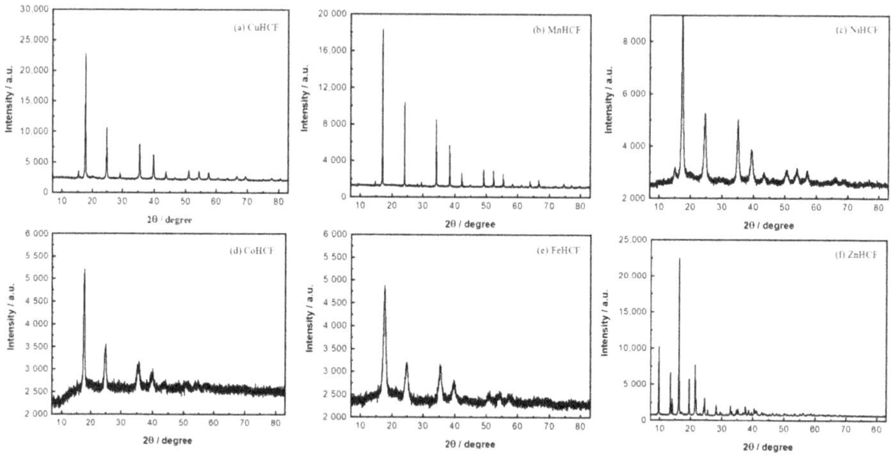

Figure 1. X-ray diffraction of PB analogues: (**a**) Cu, (**b**) Mn, (**c**) Ni, (**d**) Co, (**e**) Fe and (**f**) Zn.

Table 1. Experimental data of the crystal structure for the MHCFs.

	MnHCF	FeHCF	CoHCF	NiHCF	CuHCF	ZnHCF
R-Bragg (R_B)	2.78	0.30	4.66	1.25	2.00	3.30
Crystal system	Cubic	Cubic	Cubic	Cubic	Cubic	Trigonal
Space group	Fm-3m	Fm-3m	Fm-3m	Fm-3m	Fm-3m	R-3c
Cell Mass	1044.5	1095.2	1213.7	1238.9	1356.1	3720.4
Cell Volume (Å^3)	1156.3	1049.6	1021.8	1067.1	1035.2	4543.1
Wt%−Rietveld	100.0	100.0	100.0	100.0	100.0	100.0
a (Å)	10.492	10.106	10.097	10.219	10.107	12.648
c (Å)	–	–	–	–	–	32.647

Structural parameters (the coordinates and population of atoms, unit cell parameters and texture) were refined using the Rietveld method from the powder X-ray diffraction data (Table 1, Table 2). The structure of $Ni_4[Fe(CN)_6]_{2.67}\cdot 21H_2O$ (№151705−ICSD) was chosen as a theoretical model for MHCFs with a cubic structure. The structure of ZnHCF was refined based on the structure of $Zn_3[Fe(CN)_6]_2$ (№157850−ICSD), with a space group of R-3c.

The morphology and particle size distribution also depend on the nature of the transition metals. Scanning electron microscopy (SEM) was used to obtain images on the Hitachi S-3400N. The experimental data obtained, shown in Figure 1, are in agreement with the results of the X-ray phase analysis.

Table 2. Refinement parameters of the crystalline cell via Rietveld's method. It contains information on atomic parameters, such as coordinates and population. The value of the R_{exp} factor indicates the level of agreement between the experimental data and the calculated radiograph.

Sample	Site	Pos.	*x*	*y*	*z*	Atom	Site Population	R_{exp}	R_B
MnHCF	Mn1	4	0.5000	0.0000	0.0000	Mn^{2+}	0.852	3.16	2.78
	Fe1	4	0.0000	0.0000	0.0000	Fe^{3+}	0.666		
	C1	24	0.1880	0.0000	0.0000	C	0.666		
	N1	24	0.2997	0.0000	0.0000	N	0.666		
	O1	24	0.2889	0.0000	0.0000	O^{2-}	0.001		
	O2	32	0.1860	0.1860	0.1860	O^{2-}	0.187		
	O3	8	0.2500	0.2500	0.2500	O^{2-}	2.574		

Table 2. *Cont.*

Sample	Site	Pos.	x	y	z	Atom	Site Population	R_{exp}	R_B
FeHCF	Fe2	4	0.5000	0.0000	0.0000	Fe^{2+}	0.808	1.97	0.30
	Fe1	4	0.0000	0.0000	0.0000	Fe^{3+}	0.666		
	C1	24	0.1880	0.0000	0.0000	C	0.666		
	N1	24	0.2997	0.0000	0.0000	N	0.666		
	O1	24	0.2889	0.0000	0.0000	O^{2-}	0.002		
	O2	32	0.1860	0.1860	0.1860	O^{2-}	0.306		
	O3	8	0.2500	0.2500	0.2500	O^{2-}	0.452		
CoHCF	Co1	4	0.5000	0.0000	0.0000	Co^{2+}	0.904	2.47	4.66
	Fe1	4	0.0000	0.0000	0.0000	Fe^{3+}	0.666		
	C1	24	0.1880	0.0000	0.0000	C	0.666		
	N1	24	0.2997	0.0000	0.0000	N	0.666		
	O1	24	0.2889	0.0000	0.0000	O^{2-}	0.002		
	O2	32	0.1860	0.1860	0.1860	O^{2-}	0.298		
	O3	8	0.2500	0.2500	0.2500	O^{2-}	0.335		
NiHCF	Ni1	4	0.5000	0.0000	0.0000	Ni^{2+}	0.919	1.98	1.25
	Fe1	4	0.0000	0.0000	0.0000	Fe^{3+}	0.666		
	C1	24	0.1880	0.0000	0.0000	C	0.666		
	N1	24	0.2997	0.0000	0.0000	N	0.666		
	O1	24	0.2889	0.0000	0.0000	O^{2-}	0.027		
	O2	32	0.1860	0.1860	0.1860	O^{2-}	0.328		
	O3	8	0.2500	0.2500	0.2500	O^{2-}	0.466		
CuHCF	Cu1	4	0.5000	0.0000	0.0000	Cu^{2+}	0.906	2.50	2.00
	Fe1	4	0.0000	0.0000	0.0000	Fe^{3+}	0.666		
	C1	24	0.1880	0.0000	0.0000	C	0.666		
	N1	24	0.2997	0.0000	0.0000	N	0.666		
	O1	24	0.2889	0.0000	0.0000	O^{2-}	0.006		
	O2	32	0.1860	0.1860	0.1860	O^{2-}	0.158		
	O3	8	0.2500	0.2500	0.2500	O^{2-}	0.285		
ZnHCF	Fe1	12	0.0000	0.0000	0.1466	Fe^{3+}	1.000	3.12	3.30
	Zn1	18	0.2900	0.0000	0.2500	Zn^{2+}	1.000		
	C1	36	0.1159	−0.0150	0.1797	C	1.000		
	N1	36	0.1811	−0.0230	0.2028	N	1.000		
	C2	36	0.1302	0.1161	0.1133	C	1.000		
	N2	36	0.2163	0.1906	0.0969	N	1.000		

3.2. Morphology/SEM

SEM images were obtained, reflecting the particle morphology for all synthesized compounds, see Figure 2.

Microscopic data show that almost all samples have crystals with a distinct habit. This habit corresponds to the crystal structure clearly visible in the images. For the cubic structure of Fm-3m, crystals with a cubic shape are quite typical. It was possible to identify the crystal boundaries and analyze the particle size distribution for all the PB analogues except the ferrous one. It should be noted that depending on the divalent metal atom present in the structure, there are differences in the average particle size. In principle, all the particles of almost all the samples have sizes in the micrometre range. Only CoHCF and CuHCF substances have average particle sizes in the nanorange, measuring 50–60 nm. The particle size distribution was constructed from images. A minimum of 30 particles were selected for each compound, and the particle size distribution shown in Figure 3 was constructed.

Figure 2. *Cont.*

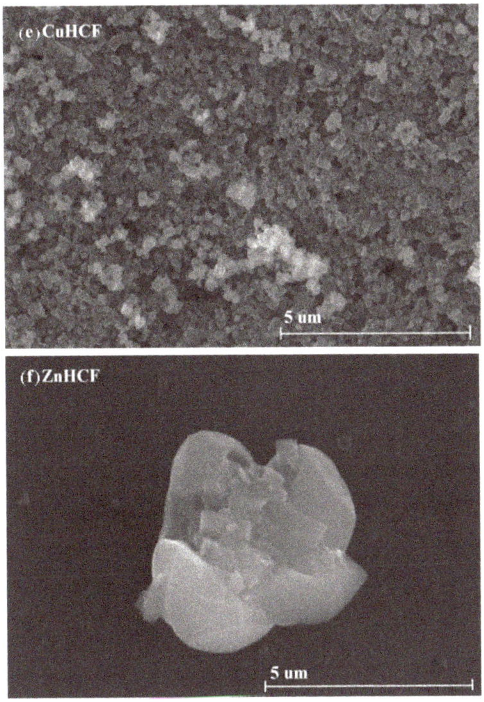

Figure 2. SEM images of PB analogues.

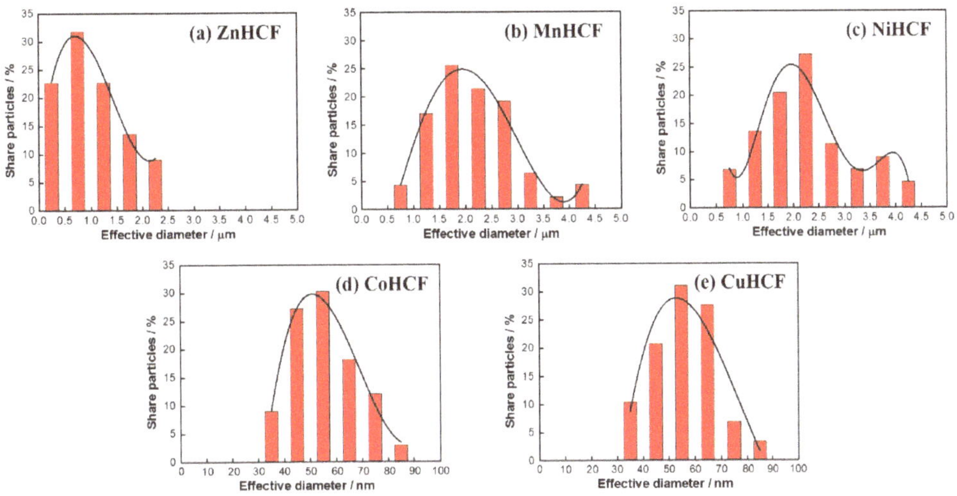

Figure 3. Particle size distribution.

3.3. Thermal Analysis

Thermal studies of the synthesized samples were carried out to determine the number of water molecules in the complexes. According to the literature, in the structure of PB analogues, it is possible that water molecules are adsorbed on the surface or intercalated into the lattice cavity, which is confirmed by the TGA-DTG diagram in Figure 4. It is possible to divide the thermal behaviour of the substances into two groups in the temperature range studied. For one group of substances (manganese, nickel and copper), two sections can be distinguished on the thermogravimetric curve. For the other substances, no such two sections can be distinguished. The different natures of the loss of mass and the temperature values are an indication of the different types of water molecules that are part of the substances. In general, the thermal stability of the compounds is similar. The same effects are observed over the temperature range studied. However, two groups of compounds can be distinguished due to some differences in the thermal dependence of the stability of these compounds. For the copper-, manganese- and nickel-containing compounds, two areas of mass decrease are clearly defined in the temperature dependence (left scale, black line) in the temperature intervals of 300–400 and 450–700 degrees Celsius. The total mass loss was not more than 35%. The same mass decrease with increasing temperature is clearly visible in the dependence of the derivative of the mass decrease (scale on the right, red line). From these dependencies it is possible to more precisely determine the value of the mass loss temperature in each temperature interval. The temperature values are 420 and 460 K (copper), 440 and 570 K (manganese) and 450 and 550 K (nickel). For the other three compounds, it is not possible to determine the exact mass loss temperature. Except for cobalt ferrocyanide, the first interval of mass loss is more pronounced, with a temperature of 430 K. The mass loss for all compounds was between 10% and not more than 40%.

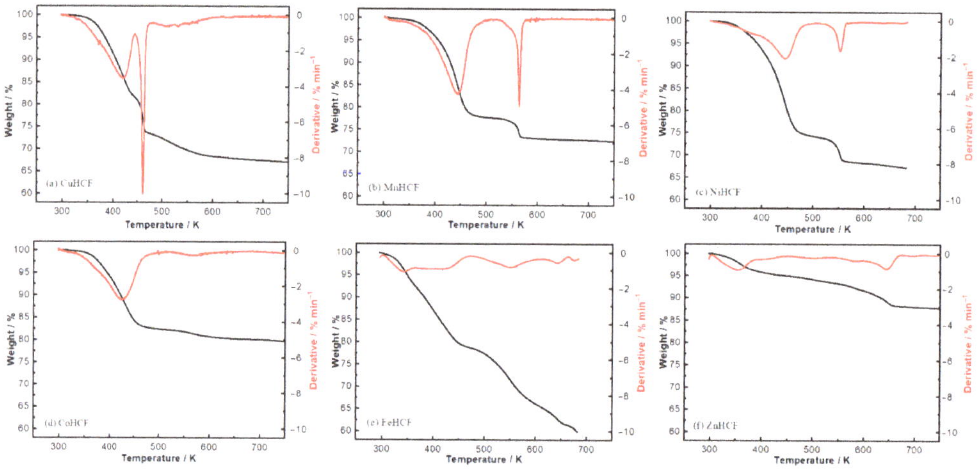

Figure 4. TGA results for the PB analogues with (**a**) Cu, (**b**) Mn, (**c**) Ni, (**d**) Co, (**e**) Fe and (**f**) Zn. The black colour lines show a decrease in the mass of the sample in percentage; the red colour lines depict the time derivative.

3.4. Chemical Analysis

An elemental analysis was carried out to determine the compound formula using the ICP. Powder samples of the PB analogues were subjected to microwave decomposition in nitric acid to obtain a solution. The mass contents of the metals and final chemical formulas are given in Tables 3 and 4.

Table 3. Results of the chemical analysis of compounds in mol.%.

Designation	K	Mn	Fe	Co	Ni	Cu	Zn
MnHCF	5.9	54.4	39.8	–	–	–	–
FeHCF	–	–	100	–	–	–	–
CoHCF	–	–	38.7	61.3	–	–	–
NiHCF	3.2	–	38.5	–	58.3	–	–
CuHCF	3.6	–	36.7	–	–	59.7	–
ZnHCF	–	–	36.3	–	–	–	63.7

Table 4. Summary data based on chemical analysis and thermal testing.

Designation	The Formula of Compound	K+ Vs. M^{2+}, mol.%	Amount of Zeolitic Water
MnHCF-1	$K_{0.21}Mn_{1.39}Fe(CN)_6{\cdot}4H_2O$	15.11 ± 0.30	1
FeHCF-1	$Fe_{1.50}Fe(CN)_6{\cdot}6H_2O$	–	–
CoHCF-1	$Co_{1.50}Fe(CN)_6{\cdot}3H_2O$	–	–
NiHCF-1	$K_{0.12}Ni_{1.44}Fe(CN)_6{\cdot}6H_2O$	8.33 ± 0.16	2
CuHCF-1	$K_{0.14}Cu_{1.43}Fe(CN)_6{\cdot}6H_2O$	9.79 ± 0.20	1
ZnHCF-1	$Zn_{1.50}Fe(CN)_6$	–	–

3.5. Electrochemical Experiments

The experimental specific capacity (Table 5, Q_{exp} aqua) was calculated by integrating the cathode branch on a cyclic voltammogram, obtained at a scanning speed of 1 mV/s in an aqueous solution for 1 M $MgSO_4$ vs. Mg^{2+}/Mg. Data have also been obtained for a cell with two terminals and a non-aqueous electrolyte (Table 5, Q_{exp} nonaqua).

Table 5. The electrochemical behaviour of the synthesized samples.

Element	E_{ox} (V)	E_{red} (V)	Q_{calc} (mAh g^{-1})	Q_{exp} (mAh g^{-1}) Aqua	Q_{exp} (mAh g^{-1}) Nonaqua	Coulomb Efficiency
CuHCF	3.45	2.88 2.51	64.3	21.0	60.8	0.913
NiHCF	3.10	2.81 2.47	65.4	19.2	63.2	0.911
MnHCF	3.15	2.84 2.67	72.7	9.8	69.2	1.015
CoHCF	3.15	2.49	75.6	6.1	72.1	0.987
FeHCF	3.47	2.63	66.3	1.5	56.7	0.952
ZnHCF	3.42	2.56	86.6	2.4	8.6	1.004

4. Discussion

The image of MnHCF shows sufficiently large crystals of a cubic shape with an average size of 2.5 ± 1.7 µm. The combination of small particles into large agglomerates with sizes ranging from 0.5 to 4.5 µm is characteristic of NiHCF. The particle sizes of CoHCF and CuHCF are in the nanometre range. Their average size is between 35 and 85 nm. The geometric parameters of the agglomerates obtained for FeHCF could not be measured. The large particles in the image are probably a cluster of smaller particles (Figure 3). As noted in the literature, e.g., [17], the preparation of Prussian Blue analogues via co-precipitation results mainly in the formation of micron-sized particles, as shown in Figure 3. In the case of copper- and cobalt-containing ferrocyanides, nanometre-sized particles can be obtained due to the different crystal growth rates. It should be noted that all substances were obtained under conditions of the same temperature, concentration and rates of addition of reagents and stirring of the solution. If growth is faster than the rate of accumulation of the critical concentration required for nucleation growth, the formation of small crystals should be expected. The explanation for the rapid growth of crystals of cobalt and copper ferricyanides may be related to the lower susceptibility of these cations to hydrolysis and consequently their greater accessibility to complexation.

Nucleation and crystallization processes play an important role in the synthesis of Prussian Blue analogues via the coprecipitation method. Differences in nucleation rates and solubility constants determine the need for the precise control of the synthesis process to obtain certain compound parameters, such as crystallinity, crystal size, defects and many others. An important factor affecting their electrochemical applications is the defects in the crystal structure of PB analogues (e.g., interstitial water molecules or vacancies). It should be noted that the composition of the cationic sublattice and the presence of zeolite water will determine the size of the resulting particles. Introducing potassium cations into the structure during synthesis is accompanied by the appearance of zeolitic water molecules. Also note that the average particle size changes with such substitution in the sublattice. The dependence of the resulting particle size on the ionic radii of the transition metals in the compound is difficult to isolate. However, it can be noted that for MnHCF, NiHCF and CuHCF, the largest difference in ionic radii of a divalent cation (Mn, Co and Cu) in an octahedral environment is observed compared to the ionic radius of low-spin trivalent iron in the same environment (Table 6). Large stresses and local distortions in the structure can result from a large difference in the ionic radii of the components. It will therefore be typical not to increase the size of the crystals but to limit their growth in compounds where large local distortions of the crystal lattice can occur. This will be accompanied by a large number of smaller-sized particles appearing. The opposite situation is observed for divalent cations. Their ionic radius differs less from the radius of low-spin trivalent iron in an octahedral environment (Fe and Ni). The situation is different for systems containing zinc, since zinc is located in a tetrahedral environment in the lattice. Zinc's ion radius in such a setting is intermediate between those of the other two cation groups.

Table 6. Comparison of polarizing ability of six coordinated metal ions in the PB analogue structure.

Ion	Spin	Crystal Radius, Å	Ionic Radius, Å	Polarizing Ability, $Å^{-1}$ Crystal	Polarizing Ability, $Å^{-1}$ Ionic	Qexp, (mA·h)/g
Mn^{2+}	Low	0.81	0.67	2.47	2.99	
	High	0.97	0.83	2.06	2.41	
Mn^{3+}	Low	0.72	0.58	4.17	5.17	10
	High	0.785	0.645	3.82	4.65	
Fe^{2+}	Low	0.75	0.61	2.67	3.28	
	High	0.92	0.78	2.17	2.56	
Fe^{3+}	Low	0.69	0.55	4.35	5.45	2
	High	0.785	0.645	3.82	4.65	
Co^{2+}	Low	0.79	0.65	2.53	3.08	
	High	0.885	0.745	2.26	2.68	
Co^{3+}	Low	0.685	0.545	4.38	5.50	6
	High	0.75	0.61	4.00	4.92	
Ni^{2+}	–	0.83	0.69	2.41	2.90	
Ni^{3+}	Low	0.7	0.56	4.29	5.36	19
	High	0.74	0.6	4.05	5.00	
Cu^{2+}	–	0.87	0.73	2.30	2.74	
Cu^{3+}	Low	0.68	0.54	4.41	5.56	21
* Zn^{2+}	–	0.88	0.74	2.27	2.70	2
Mg^{2+}	–	0.86	0.72	2.33	2.78	-
K^{+}	–	1.52	1.38	0.66	0.72	-

* tetragonal environment.

Due to the size of the zeolitic channels in the cubic crystal lattice of hexacyanoferrates, they can allocate not only alkali and alkaline earth metal cations [38,41–43] but also water molecules. According to the literature, PB analogues can contain two types of water molecules: adsorbed on the surface and intercalated in zeolitic channels [34]. As the samples dry, the adsorbed water molecules evaporate first, and then the number of intercalated molecules begins to decrease. The role of water molecules in the composition of PB analogues is still poorly understood. For example, Yang [44] states that the presence of zeolitic water has no significant effect on the electrochemical properties of cathode materials. However, in a number of other studies, researchers argue that the molecules of zeolitic water improve the diffusion kinetics by shielding the charge of the cations [45–47].

Thermal studies were carried out on SETSYS Evolution 16 in the temperature range from room temperature to 700 K. In accordance with theoretical ideas, the results obtained can be interpreted as follows. The decrease in mass of the sample in the temperature range of 413–463 K indicates the evaporation of surface adsorption water (Figure 4). Zeolitic water probably evaporates at higher temperatures from 553 to 585 K, with one mole of zeolitic water per mole of the compound in the hexacyanoferrates MnHCF and CuHCF. The amount of zeolitic water increases to 2 moles in NiHCF. Zeolitic water is absent in FeHCF, CoHCF and ZnHCF samples.

The chemical composition of the synthesized PB analogues was determined via atomic emission spectroscopy with inductively coupled plasma on the Shimadzu ICPE-9000 (Tables 3 and 4). The presence of K^+ in the hexacyanoferrates allows for the intercalation of alkali and alkaline earth metal cations into the zeolitic sites. At the same time, water molecules preferentially pass through the compounds in which K+ is absent [35,48]. In MnHCF, NiHCF and CuHCF, part of the 3d-metal cations in the sites of the cubic crystal lattice are replaced by K^+. The composition of these compounds is given by the formulae $K_{0.14}Cu_{1.43}Fe(CN)_6{\cdot}6H_2O$, $K_{0.12}Ni_{1.44}Fe(CN)_6{\cdot}6H_2O$ and $K_{0.21}Mn_{1.39}Fe(CN)_6{\cdot}4H_2O$. FeHCF, and CoHCF and ZnHCF do not contain K^+ in their composition. The molecular formula of these compounds is $Co_{1.50}Fe(CN)_6{\cdot}3H_2O$, $Fe_{1.50}Fe(CN)_6{\cdot}6H_2O$ and $Zn_{1.50}Fe(CN)_6$. The structure of the PB analogues, in which there are no alkali metal cations, becomes defective due to the presence of vacancies at the sites of the cubic crystal lattice.

The ionic radius of transition metals and the fact that zinc is no longer a 3d element may explain the difference in structural characteristics. In the case of zinc, unlike the elements of the manganese–copper series, there is a strong tendency towards tetragonal coordination. Here, the crystal structure of ZnHCF differs from all other compounds. It should be noted that it is also possible to obtain a cubic structure for zinc hexacyanoferrate. However, this can only be stabilized in the presence of zeolite water incorporated into the lattice. However, the instability of this structure for a zinc-containing compound has already been noted in [49]. Removing water molecules transforms the structure into a more stable trigonal one [50]. The cubic structure of ZnHCF has also been stabilized via syntheses in the presence of surfactants and a high content of potassium cations in the synthesis solution [51]. The variety of structural transformations is primarily related to the qualitative and quantitative composition of the cationic sublattices. Different contents of alkali metal cations and zeolites or coordinated water molecules lead to the crystallization of Prussian Blue analogues in different crystal structures [52].

The electrochemical behaviour of cathode materials in an aqueous solution of 1M $MgSO_4$ has been studied using the method of cyclic voltammetry on an Elins P-30SM potentiostat. The reaction that takes place during the intercalation of Mg^{2+} into the zeolitic channels of the cubic crystal lattice can be described as follows:

$$M_3{}^{2+}[Fe^{3+}(CN)_6]_2 + 2xe^- + xMg^{2+} = Mg_x{}^{2+}M_3{}^{2+}[Fe_{(1-x)}{}^{3+}Fe_x{}^{2+}(CN)_6]_2$$

In the course of the experiment, it was found that the second maximum in the cathode region is observed on the voltammogram for cathode materials obtained on the basis of MHCFs in the presence of K^+ (Figure 5). A similar picture is also observed for MnHCF and CuHCF. A maximum reflecting the recovery reaction can be seen on the voltammograms of FeHCF, CoHCF and ZnHCF, as shown in the example of CoHCF in Figure 5.

The appearance of a second maximum in the cathode region requires additional study of the mechanism of magnesium intercalation, which is beyond the scope of this study. According to scientists studying the intercalation of multiply charged ions in a cubic crystal lattice, the appearance of an additional wave indicates the process of desorption of an alkali metal cation during the intercalation of highly charged ions [53–56].

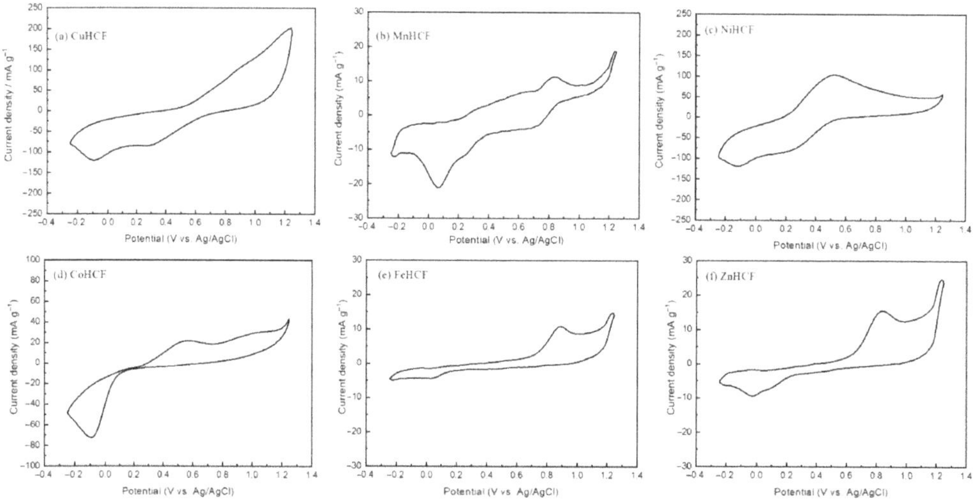

Figure 5. Cyclic voltammograms of PB analogues.

The desorption of K^+ takes place at a potential of 2.8 V for the example of the NiHCF voltammogram. The results of the electrochemical study for all synthesized MHCFs are presented in Table 6, which also includes theoretical and experimental specific capacitance values.

The theoretical specific capacitance (Q_{calc}) was calculated on the assumption that one electron participates in the reaction, according to the formula Q_{calc} = (nF)/M, where n is the number of electrons participating in the electrochemical reaction, F = 96,485.3 C/mol and is the Faraday constant and M is the molar mass of the compound in g/mol.

The lower number of electrons involved in the intercalation process is due to the large difference between the theoretical and experimental capacitance. The capacitance value can also be reduced by the presence of a part of the amorphous phase that does not participate in the electrochemical reaction. It should be noted that the samples containing potassium ions have shown a high capacitance during the experiment.

This confirms the assumption that structures of this type have a greater tendency to pass charged cations. The optimum charge/discharge current density was calculated from the cyclic voltammetry results and the capacitance values for each sample. The results of testing the samples at a constant current are shown in Figure 6. A large difference is observed for FeHCF, CoHCF and ZnHCF. A larger value is the equilibrium potential. For samples containing potassium, the value of the equilibrium potential is higher: NiHCF (8.3 mol% K vs. Ni) 2.68 V, CuHCF (9.8 mol% K vs. Cu) 2.80 V and MnHCF (15.1 mol% K vs. Mn) 2.81 V. The equilibrium potential for samples without potassium is FeHCF 2.65 V, CoHCF 2.60 V and ZnHCF 2.59 V. Similar effects were observed in [57–59]. The electrochemical capacity of the material is increased by the presence of potassium cations in the structure of cathode materials for MnHCF, NiHCF and CuHCF samples. This fact may be associated with local distortions of the structure. The larger size of the potassium cation compared to transition metal cations allows for the increase in permeability of structural channels for the process of intercalation of magnesium ions. The obtained cathode materials show a capacity close to the theoretical one in a non-aqueous medium. The stability of the charging and discharging process was studied with 60 cycles. Within the limits of this study, changes in capacity did not exceed 10%. The Coulomb efficiency is usually used to describe the released battery capacity. It refers to the ratio of the discharge capacity after the full charge and the charging capacity of the same cycle. The data are summarized in Table 5.

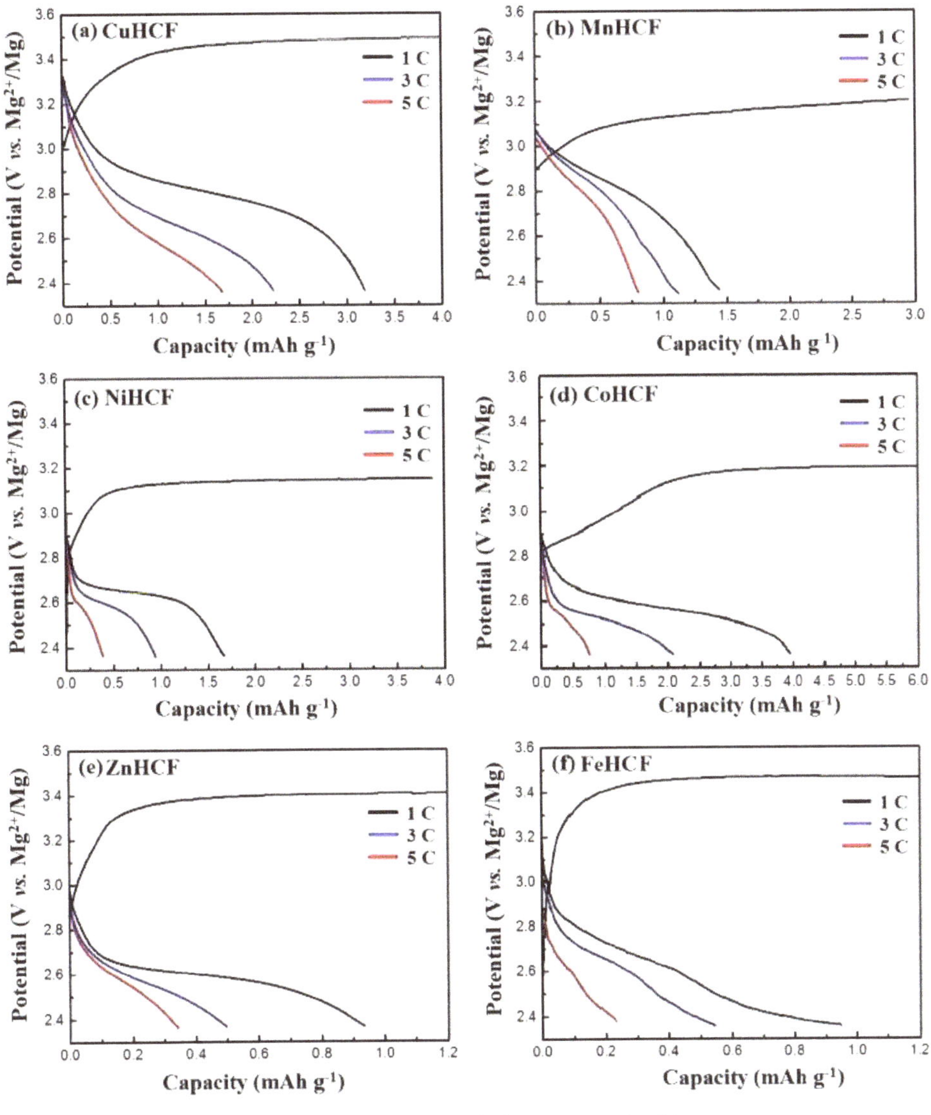

Figure 6. Charge and discharge curves at various discharge current densities.

The extensive ion channels within the PBA lattice have led to significant advances in the field of aqueous batteries. These include aqueous magnesium-ion batteries and aqueous potassium-ion batteries. However, the low decomposition voltage of H_2O [52] has severely limited the energy and power density of aqueous batteries.

5. Conclusions

In conclusion, we note that the nature of 3d-metals dramatically affects the physico-chemical properties of the PB analogues. Samples of the transition metal hexacyanoferrates MnHCF, FeHCF, CoHCF, NiHCF, CuHCF and ZnHCF have been synthesized and characterized by a wide range of techniques. The chemical composition of the samples obtained has been determined by means of atomic emission spectroscopy. X-ray phase analysis and scanning electron microscopy showed that, in spite of the same conditions for the preparation of the samples, the size and shape of the crystallites are significantly different for the different 3d-metals. The STA method revealed a different nature of mass loss of the samples. This is also dependent on the nature of the transition element in the structure. All this suggests that by introducing different transition element cations into the composition of Prussian Blue, it is possible to control the defect formation processes. A PB analogue with zinc differs from other samples due to the crystal lattice. We improve an explanation of the magnesium ion intercalation processes into PB analogue structures by taking into account not only ion size but also 3d-metals and magnesium ions. Thus, the value of the polarizing ability of magnesium ions which are surrounded by six water molecules is similar to that of 3d-metals, correlating with increased experimental capacitance magnitude (Table 6) for the samples which contain Ni and Co ions. In connection with this, it becomes urgent to investigate complexes containing magnesium ions as cathode materials for Mg-ion batteries. The study of the electrochemical characteristics of the synthesized samples of MHCFs (M = Mn, Fe, Co, Ni, Cu, Zn) showed that all the compositions studied have a low specific capacity, which makes them not very suitable to use in the pure form. However, it was found that among the compositions studied, the best properties were observed in CoHCF and NiHCF. This makes them promising not only from a fundamental point of view but also as components for the construction of hybrid cathode materials. Clear correlations between the structural and capacitive properties of the obtained cathode materials and the properties of the transition elements contained in the Prussian Blue analogues were established.

Author Contributions: Conceptualization, A.P. and A.S.; methodology, A.S.; software, S.E.; validation, A.A.; formal analysis, I.P.; investigation, S.E.; resources, A.F.; data curation, S.E.; writing—original draft preparation, A.S.; writing—review and editing, S.E.; visualization, A.A.; supervision, A.P.; project administration, A.S.; funding acquisition, A.P. All authors have read and agreed to the published version of the manuscript.

Funding: The authors acknowledge Saint Petersburg State University for research grant № 121061000160-2 that provided financial support for the development and characterization.

Data Availability Statement: Data are contained within the article.

Acknowledgments: The scientific research was performed at the Research park of St. Petersburg State University: Centre for Geo-Environmental Research and Modeling (GEOMODEL); Centre for Innovative Technologies of Composite Nanomaterials; Magnetic Resonance Research Centre; Chemical Analysis and Materials Research Centre; Centre for Optical and Laser Materials Research; Centre for Molecular and Cell Technologies; Centre for X-ray Diffraction Studies; Thermogravimetric and Calorimetric Research Centre; Centre for Physical Methods of Surface Investigation; Centre for Diagnostics of Functional Materials for Medicine, Pharmacology and Nanoelectronics; and Nanophotonics Centre.

Conflicts of Interest: The authors declare no conflicts of interest.

References

1. Krap, C.P.; Balmaseda, J.; Castillo, L.F.; Zamora, B.; Reguera, E. Hydrogen Storage in Prussian Blue Analogues: H_2 Interaction with the Metal Found at the Cavity Surface. *Energy Fuels* **2010**, *24*, 581–589. [CrossRef]
2. Kaye, S.S.; Long, J.R. Hydrogen Storage in the Dehydrated Prussian Blue Analogues $M_3[Co(CN)_6]_2$ (M = Mn, Fe, Co, Ni, Cu, Zn). *J. Am. Chem. Soc.* **2005**, *127*, 6506–6507. [CrossRef] [PubMed]
3. Matsui, J.; Kikuchi, R.; Miyashita, T. A Trilayer Film Approach to Multicolor Electrochromism. *J. Am. Chem. Soc.* **2014**, *136*, 842–845. [CrossRef] [PubMed]

4. Torad, N.L.; Hu, M.; Imura, M.; Naito, M.; Yamauchi, Y. Large Cs adsorption capability of nanostructured Prussian Blue particles with high accessible surface areas. *Y. J. Mater. Chem.* **2012**, *22*, 18261–18267. [CrossRef]
5. Aguila, D.; Prado, Y.; Koumousi, E.; Mathoniere, C.; Clerac, R. Switchable Fe/Co Prussian blue networks and molecular analogues. *Chem. Soc. Rev.* **2016**, *45*, 203–224. [CrossRef]
6. Long, J.; Guari, Y.; Guerin, C.; Larionova, J. Prussian blue type nanoparticles for biomedical applications. *Dalton Trans.* **2016**, mboxemph45, 17581–17587. [CrossRef]
7. Dharman, M.M.; Yu, J.-I.; Ahn, J.-Y.; Park, D.-W. Selective production of cyclic carbonate over polycarbonate using a double metal cyanide–quaternary ammonium salt catalyst system. *Green Chem.* **2009**, *11*, 1754–1757. [CrossRef]
8. Srivastava, R.; Srinivas, D.; Ratnasamy, P. Fe–Zn double-metal cyanide complexes as novel, solid transesterification catalysts. *J. Catal.* **2006**, *241*, 34–44. [CrossRef]
9. Zakaria, M.B.; Hu, M.; Hayashi, N.; Tsujimoto, Y.; Ishihara, S.; Imura, M.; Suzuki, N.; Huang, Y.-Y.; Sakka, Y.; Ariga, K.; et al. Thermal Conversion of Hollow Prussian Blue Nanoparticles into Nanoporous Iron Oxides with Crystallized Hematite Phase. *Eur. J. Inorg. Chem.* **2014**, *2014*, 1137–1141. [CrossRef]
10. Zhang, L.; Shi, L.; Huang, L.; Zhang, J.; Gao, R.; Zhang, D. Rational Design of High-Performance $DeNO_x$ Catalysts Based on $Mn_xCo_{3-x}O_4$ Nanocages Derived from Metal–Organic Frameworks. *ACS Catal.* **2014**, *4*, 1753–1763. [CrossRef]
11. Wessells, C.D.; Peddada, S.V.; McDowell, M.T.; Huggins, R.A.; Cui, Y. The Effect of Insertion Species on Nanostructured Open Framework Hexacyanoferrate Battery Electrodes. *J. Electrochem. Soc.* **2012**, *159*, A98–A103. [CrossRef]
12. Omarova, M.; Koishybay, A.; Yesibolati, M.; Mentbayeva, A.; Umirov, N.; Ismailov, K.; Adair, D.; Baba, M.; Kurmanbayeva, I.; Bakenov, Z. Nickel Hexacyanoferrate Nanoparticles as a Low Cost Cathode Material for Lithium-Ion Batteries. *Electrochim. Acta* **2015**, *184*, 58–63. [CrossRef]
13. Wang, R.; Shyam, B.; Stone, K.; Weker, J.; Pasta, M.; Lee, H.; Toney, M.; Cui, Y. Reversible Multivalent (Monovalent, Divalent, Trivalent) Ion Insertion in Open Framework Materials. *Adv. Energy Mater.* **2015**, *1401869*, 1–10. [CrossRef]
14. Wang, Q.; Li, J.; Jin, H.; Xin, S.; Gao, H. Prussian-blue materials: Revealing new opportunities for rechargeable batteries. *InfoMat.* **2022**, *4*, e12311. [CrossRef]
15. Wessells, C.; Huggins, R.; Cui, Y. Copper hexacyanoferrate battery electrodes with long cycle life and high power. *Nat. com.* **2011**, *2*, 1–5. [CrossRef]
16. Sun, X.; Ji, X.; Zhou, Y.; Shao, Y.; Zang, Y.; Wen, Z.; Chen, C. A new gridding cyanoferrate anode material for lithium and sodium ion batteries: $Ti_{0.75}Fe_{0.25}[Fe(CN)_6]_{0.96}\cdot 1.9H_2O$ with excellent electrochemical properties. *J. Power Sources* **2016**, *314*, 35–38. [CrossRef]
17. Zhang, Q.; Fu, L.; Luan, J.; Huang, X.; Tang, Y.; Xie, H.; Wang, H. Surface engineering induced core-shell Prussian blue@polyaniline nanocubes as a high-rate and long-life sodium-ion battery cathode. *J. Power Sources* **2018**, *395*, 305–313. [CrossRef]
18. Ke, X.; Zhao, Z.; Liu, J.; Shi, Z.; Li, Y.; Zhang, L.; Zhang, H.; Chen, Y.; Guo, Z.; Wu, Q.; et al. Improvement in capacity retention of cathode material for high power density lithium ion batteries: The route of surface coating. *Appl. Energy* **2017**, *194*, 540–548. [CrossRef]
19. Yabuuchi, N.; Kubota, K.; Dahbi, M.; Komaba, S. Research Development on Sodium-Ion Batteries. *Chem. Rev.* **2014**, *114*, 11637. [CrossRef]
20. Yoo, H.; Shterenberg, I.; Gofer, Y.; Gershinsky, G.; Pour, N.; Aurbach, D. Mg rechargeable batteries: An on-going challenge. *Energy Environ. Sci.* **2013**, *8*, 2265–2279. [CrossRef]
21. Li, C.; Lin, L.; Wu, W.; Sun, X. A High Potential Polyanion Cathode Material for Rechargeable Mg-Ion Batteries. *Small Methods* **2022**, *6*, 2200363. [CrossRef] [PubMed]
22. Kuperman, N.; Cairns, A.; Goncher, G.; Solanki, R. Structural water enhanced intercalation of magnesium ions in copper hexacyanoferrate nonaqueous batteries. *Electrochim. Acta* **2020**, *362*, 137077. [CrossRef]
23. Matthew, M.; Bock, D.; Takeuchi, E.; Marschilok, A.; Takeuchi, K. Cathode materials for magnesium and magnesium-ion based batteries. *Coord. Chem. Rev.* **2015**, *287*, 15–27.
24. Levi, E.; Mitelman, A.; Aurbach, D.; Brunelli, M. Structural Mechanism of the Phase Transitions in the Mg−Cu−Mo6S8 System Probed by ex Situ Synchrotron X-ray Diffraction. *Chem. Mater.* **2007**, *19*, 5131–5142. [CrossRef]
25. Levi, E.; Gershinsky, G.; Aurbach, D.; Isnard, O. Crystallography of Chevrel Phases, MMo_6T_8 (M = Cd, Na, Mn, and Zn, T = S, Se) and Their Cation Mobility. *Inorg. Chem.* **2009**, *48*, 8751–8758. [CrossRef]
26. Levi, E.; Gofer, Y.; Aurbach, D. On the Way to Rechargeable Mg Batteries: The Challenge of New Cathode Materials. *Chem. Mater.* **2010**, *22*, 860–868. [CrossRef]
27. Peng, B.; Liang, J.; Tao, Z.; Chen, J. Magnesium nanostructures for energy storage and conversion. *J. Mater. Chem.* **2009**, *19*, 2877–2883. [CrossRef]
28. Wessells, C.; McDowell, M.; Peddada, S.; Pasta, M.; Huggins, R.; Cui, Y. Tunable Reaction Potentials in Open Framework Nanoparticle Battery Electrodes for Grid-Scale Energy Storage. *ACS Nano* **2012**, *6*, 1688–1694. [CrossRef]

29. Yang, D.; Xu, J.; Liao, X.-Z.; Wang, H.; He, Y.-S.; Ma, Z.-F. Retracted Article: Prussian blue without coordinated water as a superior cathode for sodium-ion batteries. *Chem. Com.* **2015**, *51*, 8181–8184. [CrossRef]
30. Liu, S.; Pan, G.; Li, G.; Gao, X.P. Copper hexacyanoferrate nanoparticles as cathode material for aqueous Al-ion batteries. *J. Mat. Chem. A* **2015**, *3*, 959–962. [CrossRef]
31. Lipson, A.L.; Han, S.; Kim, S.; Pan, B.; Sa, N.; Liao, C.; Fister, T.T.; Burrell, A.K.; Vaughey, J.T.; Ingram, B.J. Nickel hexacyanoferrate, a versatile intercalation host for divalent ions from nonaqueous electrolytes. *J. Power Sources* **2016**, *325*, 646. [CrossRef]
32. Marzak, P.; Kosiahn, M.; Yun, J.; Bandarenka, A.S. Intercalation of Mg^{2+} into electrodeposited Prussian Blue Analogue thin films from aqueous electrolytes. *Electrochim. Acta* **2019**, *307*, 157. [CrossRef]
33. Kim, D.-M.; Kim, Y.; Arumugam, D.; Won Woo, S.; Jo, Y.N.; Park, M.-S.; Kim, Y.-J.; Choi, N.-S.; Lee, K.T. Co-intercalation of Mg^{2+} and Na^+ in $Na_{0.69}Fe_2(CN)_6$ as a High-Voltage Cathode for Magnesium Batteries. *ACS Appl. Mater. Interfaces* **2016**, *8*, 8554–8560. [CrossRef] [PubMed]
34. Yagi, S.; Fukuda, M.; Ichitsubo, T.; Nitta, K.; Mizumaki, M.; Matsubara, E. EQCM Analysis of Redox Behavior of CuFe Prussian Blue Analog in Mg Battery Electrolytes. *J. Electrochem. Soc.* **2015**, *162*, A2556. [CrossRef]
35. Chen, L.; Bao, J.L.; Dong, X.; Truhlar, D.G.; Wang, Y.; Wang, C.; Xia, Y. Aqueous Mg-Ion Battery Based on Polyimide Anode and Prussian Blue Cathode. *ACS Energy Lett.* **2017**, *2*, 1115–1121. [CrossRef]
36. Mizuno, Y.; Okubo, M.; Hosono, E.; Kudo, T.; Oh-ishi, K.; Okazawa, A.; Kojima, N.; Kurono, R.; Nishimura, S.; Yamada, A. Electrochemical Mg^{2+} intercalation into a bimetallic CuFe Prussian blue analog in aqueous electrolytes. *J. Mater. Chem. A* **2013**, *1*, 13055–13059. [CrossRef]
37. Wang, R.; Wessells, C.; Huggins, R.; Cui, Y. Highly Reversible Open Framework Nanoscale Electrodes for Divalent Ion Batteries. *Nano Lett.* **2013**, *13*, 5748–5752. [CrossRef]
38. Mizuno, Y.; Okubo, M.; Hosono, E.; Kudo, T.; Zhou, H.; Oh-ishi, K. Suppressed Activation Energy for Interfacial Charge Transfer of a Prussian Blue Analog Thin Film Electrode with Hydrated Ions (Li^+, Na^+, and Mg^{2+}). *J. Phys. Chem. C* **2013**, *117*, 10877–10882. [CrossRef]
39. Robin, M.; Day, P. Mixed Valence Chemistry-A Survey and Classification. *Adv. Inorg. Chem. Radiochem.* **1968**, *10*, 247–422.
40. Li, Q.; Kharzeev, D.; Zhang, C.; Huang, Y.; Pletikosić, I.; Fedorov, A.V.; Zhong, R.D.; Schneeloch, J.A.; Gu, G.D.; Valla, T. Chiral magnetic effect in $ZrTe_5$. *Nature Phys* **2016**, *12*, 550–554. [CrossRef]
41. Ling, C.; Chen, J.; Mizuno, F. First-principles study of alkali and alkaline earth ion intercalation in iron hexacyanoferrate: The important role of ionic radius. *J. Phys. Chem. C.* **2013**, *117*, 21158–21165. [CrossRef]
42. Lipson, A.; Pan, B.; Lapidus, S.; Liao, C.; Vaughey, J.; Ingram, B. Rechargeable Ca-ion batteries: A new energy storage system. *Chem. Mater.* **2015**, *27*, 8442–8447. [CrossRef]
43. Imanishi, N.; Morikawa, T.; Kondo, J.; Takeda, Y.; Yamamoto, O.; Kinugasaand, N.; Yamagishi, T. Lithium intercalation behavior into iron cyanide complex as positive electrode of lithium secondary battery. *J. Pow. Sourc.* **1999**, *79*, 215–219. [CrossRef]
44. Kim, H.; Hong, J.; Park, K.-Y.; Kim, H.; Kim, S.-W.; Kang, K. Aqueous rechargeable Li and Na ion batteries. *Chem. Rev.* **2014**, *114*, 11788–11827. [CrossRef]
45. Padigi, P.; Thiebes, J.; Swan, M.; Goncher, G.; Evans, D.; Solanki, R. Prussian Green: A high rate capacity cathode for potassium ion batteries. *Electrochim. Acta.* **2015**, *166*, 32–39. [CrossRef]
46. Nie, P.; Shen, L.; Luo, H.; Ding, B.; Xu, G.; Wang, J.; Zhang, X. Prussian blue analogues: A new class of anode materials for lithium ion batteries. *J. Mater. Chem. A.* **2014**, *2*, 5852–5857. [CrossRef]
47. Christensen, P.; Hamnett, A.; Higgins, S. A study of electrochemically grown Prussian blue films using Fourier-transform infra-red spectroscopy. *J. Chem. Soc. Dalton Trans.* **1990**, *9*, 2233–2238. [CrossRef]
48. Eftekhari, A. Potassium secondary cell based on Prussian blue cathode. *J. Power Sources* **2004**, *126*, 221–228. [CrossRef]
49. Gravereau, P.; Garnier, E. Structure de la phase cubique de l'hexacyanoferrate(III) de zinc: $Zn_3[Fe(CN)_6]_2.nH_2O$. *Acta Cryst.* **1984**, *C40*, 1306–1309. [CrossRef]
50. Du, C.-J.; Bu, F.-X.; Jiang, D.-M.; Zhang, Q.-H.; Jiang, J.-S. Prussian blue analogue $K_2Zn_3[Fe(CN)_6]_2$ quasi square microplates: Large-scale synthesis and their thermal conversion into a magnetic nanoporous $ZnFe_{2-x}O_4$–ZnO composite. *CrystEngComm* **2013**, *15*, 10597–10603. [CrossRef]
51. Yuan, Y.; Xu, J.; Li, X.; Li, X.; Wang, X.; Li, X.; Dai, S. High-quality Prussian blue analogues $K_2Zn_3[Fe(CN)_6]_2$ crystals as a stable and high rate cathode material for potassium-ion batteries. *J. Alloys Comp.* **2022**, *923*, 166457. [CrossRef]
52. Fayaz, M.; Lai, W.; Li, J.; Chen, W.; Luo, X.; Wang, Z.; Chen, Y.; Li, D.; Abbas, S.M.; Chen, Y. Prussian blue analogues and their derived materials for electrochemical energy storage: Promises and Challenges. *Mat. Res. Bull.* **2024**, *170*, 112593. [CrossRef]
53. Yagi, S.; Fukuda, M.; Makiura, R.; Ichitsubo, T.; Matsubara, E. EQCM analysis of redox behavior of Prussian blue in a lithium battery electrolyte. *J. Mater. Chem. A.* **2014**, *2*, 8041–8047. [CrossRef]
54. Trocoli, R.; Mantia, F.L. An Aqueous Zinc-ion battery based on copper hexacyanoferrate. *ChemSusChem* **2015**, *8*, 481–485. [CrossRef] [PubMed]

55. Reed, L.; Ortiz, S.; Xiong, M.; Menke, E. A rechargeable aluminum-ion battery utilizing a copper hexacyanoferrate cathode in an organic electrolyte. *Chem. Commun.* **2015**, *51*, 14397–14400. [CrossRef] [PubMed]
56. Bock, C.; Kaufman, A.; Glusker, J. Coordination of water to magnesium cations. *Inorg. Chem.* **1994**, *33*, 419–427. [CrossRef]
57. Sun, X.; Duffort, V.; Nazar, L.F. Prussian Blue Mg—Li Hybrid Batteries. *Adv. Sci.* **2016**, *3*, 1600044. [CrossRef]
58. Wang, F.; Xiulin Fan, X.; Gao, T.; Sun, W.; Ma, Z.; Yang, C.; Han, F.; Xu, K.; Wang, C. High-Voltage Aqueous Magnesium Ion Batteries. *ACS Cent. Sci.* **2017**, *3*, 1121–1128. [CrossRef]
59. Zheng, J.; Guo, G.; Li, H.; Wang, L.; Wang, B.; Yu, H.; Yan, Y.; Yang, D.; Dong, A. Elaborately Designed Micro–Mesoporous Graphitic Carbon Spheres as Efficient Polysulfide Reservoir for Lithium–Sulfur Batteries. *ACS Energy Lett.* **2017**, *2*, 1105–1114. [CrossRef]

energies

MDPI

Article

Protective Coating for Zinc Electrodes of Zinc–Air Battery in a Neutral Electrolyte

Sonia Bagheri [1,*], Benedetto Bozzini [2], Carola Esposito Corcione [1], Raffaella Striani [1] and Claudio Mele [1,*]

1 Department of Engineering for Innovation, University of Salento, Via Monteroni, 73100 Lecce, Italy; carola.corcione@unisalento.it (C.E.C.); raffaella.striani@unisalento.it (R.S.)

2 Department of Energy, Politecnico di Milano, Via Lambuschini 4, 20156 Milano, Italy; benedetto.bozzini@polimi.it

* Correspondence: sonia.bagheri@unisalento.it (S.B.); claudio.mele@unisalento.it (C.M.)

Abstract

This work introduces a novel approach to enhancing the performance of zinc anodes in zinc–air batteries through a photopolymerizable organic–inorganic hybrid coating. Electrochemical tests were conducted in a neutral NaCl electrolyte, selected to minimize electrolyte carbonation, anode corrosion, and zinc dendrite formation. The behavior of bare and coated zinc electrodes was investigated using linear sweep voltammetry, electrochemical impedance spectroscopy (EIS), potentiostatic measurements, galvanostatic discharge tests, and charge-discharge tests, while morphological and structural characterizations were carried out by Atomic Force Microscopy (AFM), Raman spectroscopy, and X-ray Diffraction (XRD). The results confirmed that the hybrid coating acts as a corrosion-resistant barrier, enhancing the reversibility and stability of zinc electrodes through a barrier mechanism. Charge–discharge tests further confirmed the improved performance of the coated electrode, obtaining at a current density of 1 mA/cm^2, a coulombic efficiency of 92.61% and a capacity retention of 90.18%, respectively, after 16 cycles. These findings highlight the effectiveness of the photopolymerizable hybrid coating in improving the durability and rechargeability of zinc–air batteries.

Keywords: photopolymerizable coating; zinc electrode; AFM; Zn-air battery; neutral electrolyte

1. Introduction

Rechargeable batteries operate as electrochemical energy storage systems by transforming chemical energy into electrical energy through oxidation-reduction reactions occurring between the anode and the cathode [1,2]. Among various rechargeable batteries, metal–air batteries (MABs) have gained attention as promising systems because of their cost-effective anode materials, such as aluminum, zinc, and iron, and atmospheric oxygen as the cathodic reactant. This configuration reduces the overall cost and significantly decreases battery weight [3–6]. Zinc-air batteries (ZABs), in particular, have gained widespread attention. Zinc offers a high theoretical specific capacity (820 mAh g^{-1} and 5855 Ah L^{-1}) and a low electrochemical potential (−0.762 V vs. SHE), making ZABs well-suited for large-scale applications. Their further advantages include high energy density, natural abundance, low toxicity, affordability, excellent safety, long shelf life (under dry conditions), environmental friendliness, and stable performance across a range of temperatures and loads [7–14].

Despite these benefits, ZABs face several challenges: limited power output, irreversibility of redox reactions at the air electrode, dendrite formation, hydrogen evolution due to anode corrosion, sluggish oxygen reduction kinetics, narrow temperature operating ranges, carbonation of alkaline electrolytes, and high overpotentials at the air cathode [10–12,15].

Some of these issues are mainly related to alkaline Zn-air batteries, where the high concentration of OH^- ions promotes dendrite growth, air cathode corrosion, and electrolyte carbonation. Neutral electrolytes have been proposed as an alternative to overcome these limitations. Lower OH^- concentrations help suppress carbonation, reduce electrode corrosion, and inhibit dendrite formation, thereby extending battery cycle life. Neutral or near-neutral electrolytes also offer enhanced safety and non-toxicity, making them attractive for practical ZAB applications [16–18].

The use of chloride-based neutral electrolytes was first introduced in 1973 for primary ZABs [19], and later extended to rechargeable systems in 2012 [17,20]. Since then, Cl^--based neutral electrolytes have been widely studied [16,20–23]. However, they can induce chlorine evolution reactions, leading to a pH drop in the electrolyte. To mitigate this, NaCl-based solutions have been recommended for better pH control, improved ionic conductivity, and minimized chlorine evolution [18].

Several studies have investigated NaCl-based systems for Zn-air batteries. The performance of zinc and Zn-Al alloy anodes has been examined in NaCl electrolytes [16]; Zinc electrode behavior has been analyzed in pure and EDTA-modified NaCl solutions [17]. Seawater-based electrolytes have been assessed to evaluate the impact of chloride ions on cathode electrocatalytic activity and overall ZAB performance [24]. Despite these advances, challenges remain, and further research is essential to fully realize the potential of NaCl-based neutral electrolytes in Zn-air battery systems [25–27].

In addition to electrolyte optimization, considerable research has been devoted to developing interfacial strategies aimed at mitigating the intrinsic challenges of zinc anodes. Among these, surface coatings have emerged as one of the most effective approaches, as they can promote uniform zinc deposition while suppressing undesirable side reactions. Because organic coatings alone are susceptible to fracture after several charge/discharge cycles, organic–inorganic hybrid coatings have shown significant promise. These systems combine the advantageous properties of both organic and inorganic components: the organic phase provides corrosion resistance, flexibility, and excellent barrier properties, whereas the inorganic phase enhances zinc-ion desolvation and facilitates faster ion transport kinetics. Moreover, the use of photopolymerization enables controlled, light-induced curing, resulting in the rapid formation of dense, homogeneous, and defect-free protective layers on the zinc surface. Furthermore, photopolymerization helps address the environmental issues associated with polymers in solutions, as it significantly reduces volatile solvent emissions. This synergistic design not only stabilizes the anode/electrolyte interface but also prolongs the electrochemical stability and cycle life of zinc-based batteries [28–30].

This study focused on assessing the role of a photopolymerizable organic-inorganic coating, labelled as "Hybrid", on the performance of zinc anode material within a Zinc-air battery configuration, all set within a 1 M NaCl solution, to evaluate the potential advantages for use in a rechargeable battery. Its specially formulated, light-curing, solvent-free coating is known for its nanostructured composition and exceptional protective properties. The coating was designed to protect porous substrates, such as natural stone, and due to the proven effectiveness of Hybrid coating in reducing the corrosion rate, its potential as an anti-corrosive top-coating for non-porous substrates, including metals such as pure zinc, is explored, which was the focus of our investigation. Hybrid is characterized by its super-hydrophobic nature, making it an ideal candidate for providing long-lasting protection to the surfaces [31,32]. The choice of a photopolymerizable system is related to the benefits of the photopolymerization process. For several years, it has been well known that light-curing is a valid alternative to traditional thermal curing, as it is a solvent-free technique that is usually carried out at room temperature in a very short time (within a few hours). Consequently, photo-cured polymeric films demonstrate enhanced performance in terms of hydrophobicity, thermal-mechanical properties, chemical resistance and durability compared to coatings produced from conventional polymeric solutions or emulsions [33,34].

In this study, the efficacy of this innovative coating has been evaluated through electrochemical and spectroelectrochemical experiments. Finally, the general discharge performance of the zinc-air battery is carefully assessed through long-term discharge tests and a series of charge/discharge tests at various current densities.

2. Materials and Methods

2.1. Materials

For the electrochemical characterization tests and morphological measurements, the electrodes were plates of 99% pure zinc (E-Nettech, Mumbai, India). For all tests, the samples had an exposed surface area of 1.44 cm^2 in contact with the electrolyte, except for the charge and discharge test, where it was 3.30 cm^2. To prepare the nearly neutral solution (1 M NaCl, pH 7), analytical grade NaCl (Sigma-Aldrich, Darmstadt, Germany) and distilled water were used. The volume of electrolyte for the electrochemical measurements was 200 mL, and for the charge/discharge test, it was approximately 5 mL.

The coating used in this study is a photopolymerizable product developed by some of the authors as a protective coating for several surfaces. The synthesis procedure and the composition of the hybrid coating are covered by a European patent [35]. The formulation is crosslinkable by solar radiations, according to the invention, contains at least 4–15% by weight of an alkoxy- silane compound, 22–90% by weight of a trimethoxypropyl silane methacrylate and functionalized poly(dimethylsiloxane)-terminated vinyl, 5–60% by weight of a methacrylic resin and 1–5% by weight of a photoinitiator mixture containing at least one UV photoinitiator and at least one photoinitiator that can be activated by visible light. The amounts given in % by weight are referred to the total weight of the organic-inorganic hybrid formulation [35].

The application of the hybrid coating involved the following steps:

(i) Immersion: the bare samples were immersed in the hybrid coating solution. The immersion was total or partial, depending on the specific test to be conducted. The samples were immersed for 6 min, with each face being immersed for approximately 3 min. This step ensured that the hybrid coating covered the entire surface of the samples.

(ii) Exposure to light radiation: after the immersion step, the samples were exposed to light radiation. This exposure was achieved by placing the samples under natural sunlight or using a sunlamp for 10 h. This step was crucial for the light-curing process of the hybrid coating, which enhances its stability and durability.

(iii) Dual-curing: following the light exposure, the samples underwent a dual-curing process. The samples were subjected to heating in an oven at a temperature of 120 °C for 1 h. This step enabled the complete curing and bonding of the hybrid coating to the surface of the samples. The dual-curing process ensured optimal adhesion.

2.2. Electrochemical, Morphological and Spectroscopic Tests

To assess the effectiveness of the hybrid coating on the corrosion resistance of zinc samples, various electrochemical tests were performed at room temperature employing a Parstat 2273 potentiostat/galvanostat (Ametek SI, Oak Ridge, TN, USA) with an integrated Impedance Analyzer linked directly to the cell.

The electrochemical cell used for the electrochemical tests was composed of a conventional three-electrode configuration. The investigated working electrodes were bare and Hybrid-coated zinc samples. A platinized titanium mesh functioned as the counter electrode, with Ag/AgCl (KCl 3 M) as the reference electrode. All measured potentials are referenced to Ag/AgCl. The electrochemical characterization was done by performing open circuit potential (OCP), linear sweep voltammetry (LSV), potentiostatic (PS), electrochemical impedance spectroscopy (EIS), galvanostatic discharge and charge/discharge tests.

OCP was recorded for 30 min, verifying that the voltage fluctuations were less than 5 mV. LSV tests were performed at a scan rate of 1 mV/s. Corrosion current density (i_{corr}) and corrosion potential (E_{corr}) were determined from LSV curves by Origin software (version 8.5). The potentiostatic corrosion tests were conducted at a constant potential of −0.8 V for 60 min. The EIS measurements were carried out on bare and Hybrid-coated samples, in the frequency range from 1 MHz to 10 mHz, at OCP with an AC amplitude of 10 mV. To obtain the related corrosion parameters, an equivalent circuit model was applied. Each experiment was conducted in triplicate to guarantee repeatability.

Galvanostatic discharge and charge/discharge tests were performed in a self-fabricated laboratory plane-parallel primary battery [13,36] composed of a cathode (a MnO_2-carbon black (Vulcan) air cathode on a Ni grid), a bare zinc disc, and the Hybrid-coated zinc serving as the anode electrode and a Celgard 3510 separator. This setup also featured a Ni plate functioning as the anode current collector. A sketch of the Zn-air battery's components is presented in Figure 1. The discharge capacitance of the Zn-air battery with bare zinc and Hybrid-coated zinc as anode material was evaluated at a fixed current of 1 mA/cm^2 to a cut-off voltage of 0.5 V and the charge/and discharge tests were conducted at the current densities of 1 mA/cm^2 and 10 mA/cm^2 for 24 h.

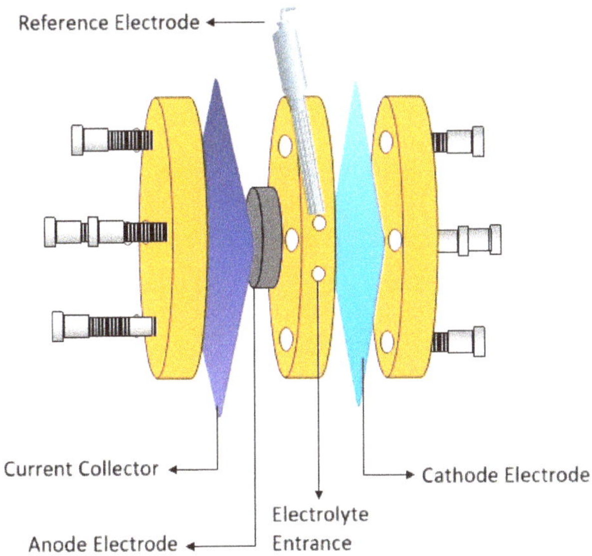

Figure 1. Graphical sketch of the zinc-air battery's components, including the zinc anode, air cathode, electrolyte and separator.

Notably, for each test iteration, a fresh cathode material was employed. Charge/discharge tests were performed based on a half-cell assembly with a zinc-base sample, with and without coating, as an anode and cathode and a zinc wire as a reference electrode.

Surface roughness and topography of the samples were measured in air employing a Bruker Multimode 8 AFM operating in quantitative nanomechanical mode, providing valuable height and peak force error data. The scan size of the images was 3 × 3 μm^2 and the scan rate was 0.5 Hz. The cantilever was RTESPA-300 (Bruker, Dresden, Germany), having a resonance frequency of about 300 kHz and a spring constant of 40 N/m. For the data processing, Nanoscope Analysis software (version 1.5) was applied.

The chemical nature of the samples, with and without coating, was studied before and after corrosion experiments via Raman spectroscopy using a LabRam confocal microscope (Palaiseau, France), a 10× objective, and 633 nm He–Ne laser excitation at 7 mW.

XRD analyses were carried out using a Rigaku Ultima+ diffractometer equipped with a Bragg–Brentano goniometer (Rigaku, Neu-Isenburg, Germany) and Cu Kα radiation (λ = 1.5406 Å), at 40 kV and 20 mA, with a 0.02° step size.

3. Results

3.1. Electrochemical Tests

The effectiveness of the hybrid coating has been evaluated by subjecting both bare and Hybrid-coated zinc samples to a series of electrochemical experiments within a 1M NaCl electrolyte. The outcomes of each of these experiments are discussed in the subsequent sections. The average coating thickness was measured by a digital micrometer (Mcbazel, Wenzhou, Zhejiang, China) and was 130 ± 6 μm, confirming a homogeneous layer formation.

OCP values of the samples were recorded by immersing them in the electrolyte for 30 min and monitoring the potential change until its fluctuations reached less than 5 mV. For the uncoated zinc sample, immediately after immersion, the potential showed a slight decrease, and approximately 15 min after immersion, it reached an almost constant value. Contrarily, the potential profiles of the Hybrid-coated zinc remained notably stable throughout the immersion duration, sustaining this stability during the whole test. At the end of the test, the open-circuit potentials for the zinc and Hybrid-coated zinc samples were recorded as −1.03 and −1.01 V, respectively. The lower activity of the Hybrid-coated sample compared to the bare sample suggests the formation of a barrier film between the substrate and the electrolyte.

The linear sweep voltammetry (LSV) curves for both bare and Hybrid-coated zinc samples are presented in Figure 2A. The LSV results show that, beyond the activation region, the current density remained nearly constant with increasing potential, suggesting the formation of a surface oxide layer [23,37]. As anticipated, the bare zinc exhibited a more negative E_{corr} and a significantly higher i_{corr} than the coated sample, highlighting the enhanced corrosion resistance of the Hybrid-coated sample. Quantitative values of the polarization resistance (Rp) of the samples were calculated from corrosion potential and current and also the Tafel slopes, which were found to be 0.119 and 0.681 $\Omega.cm^2$ for bare and Hybrid-coated zinc, respectively. The results imply that the hybrid protective coating can substantially reduce the corrosion rate of zinc [38].

The effect of the applied coating was also evaluated by potentiostatic tests carried out for 60 min at a selected potential of −0.8 V, a value corresponding to an almost constant current density in the LSVs. The graphs of the measured current density over time are shown in Figure 2B. The samples demonstrated a rise in current density initially, which subsequently stabilized, reaching nearly constant values of 42 and 14 mA/cm^2 after 60 min for zinc without and with hybrid coating, respectively. The reduced current density observed for the coated zinc revealed the satisfying corrosion behaviour of Hybrid as a protective coating against corrosion.

The electrochemical behavior and corrosion resistance of the Hybrid-coated samples were further investigated by electrochemical impedance spectroscopy (EIS). As shown in the Nyquist plots (Figure 3A), the capacitive loop exhibited a noticeably larger diameter, indicating enhanced corrosion resistance of the coated sample, which demonstrates the effectiveness of the coating [21,39].

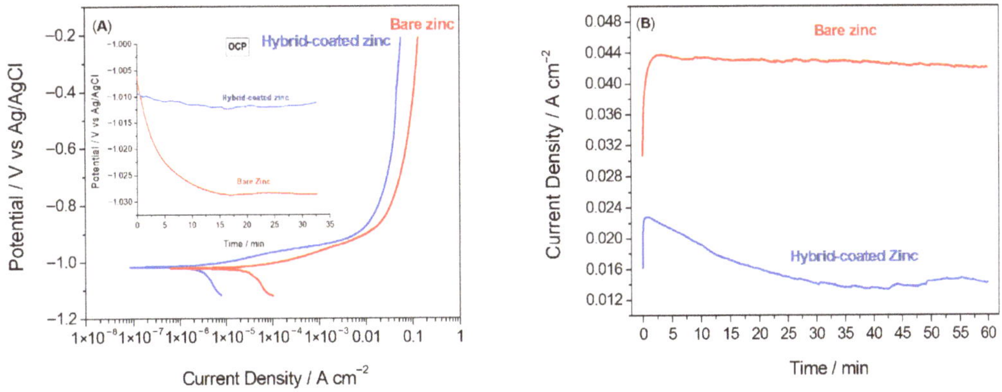

Figure 2. (**A**) Linear sweep voltammetry curves of bare and Hybrid-coated zinc samples (Inset: Open circuit potential of the bare and Hybrid-coated zinc vs. time), (**B**) at −0.8 V on bare versus Hybrid-coated zinc samples in 1 M NaCl electrolyte.

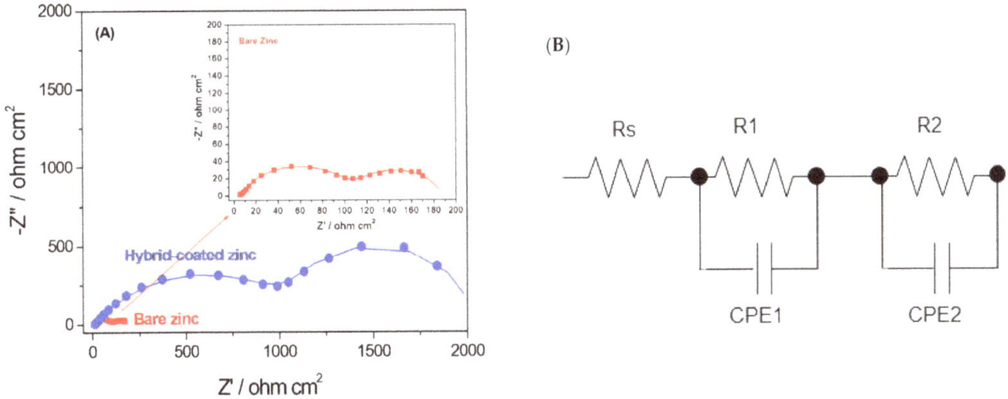

Figure 3. (**A**) Nyquist plots for bare and Hybrid-coated zinc, showing the experimental EIS data (symbols) and the corresponding fit results (lines), (**B**) schematic of the equivalent circuit used to fit the EIS data.

Two capacitive time constants were evident in the Nyquist plots of the coated and bare zinc samples. For bare zinc, the presence of two capacitive time constants indicated that the corrosion products formed on its surface were composed of two relatively dense and porous layers. [39] The equivalent circuit used to analyse the EIS data and fit the plots is shown in Figure 3B. Rs represents the electrolyte resistance. The high-frequency capacitive time constant corresponds to the interfacial film resistance (R_1), reflecting the resistance of corrosion products created on the surface of the sample. The low-frequency capacitive time constant is associated with the charge transfer resistance (R_2) at the electrode/electrolyte interface, which depends on surface conditions and indicates the resistance to charge transfer during electrochemical reactions [39,40]. CPE1 represents the coating capacitor, which is associated with the dielectric character of the coating material. It reflects the capacitive behaviour of the protective layer. CPE2 is the double-layer capacitor, which is attributed to the capacitance at the electrode/electrolyte interface. Y represents the magnitude of the constant phase element (CPE), and n is the exponent that reflects the deviation from ideal capacitive behavior. When n = 1, the CPE functions as an ideal capacitor, whereas n = 0 corresponds to a pure resistor; and when n = 0.5, it corresponds to a Warburg element, reflecting diffusion-controlled processes [39,41–43].

The resulting values of the fits are shown in Table 1. In particular, the R_1 value, representing the coating resistance, was significantly higher for the Hybrid-coated zinc compared to the bare sample. This outcome confirms the effectiveness of the hybrid coating in enhancing the corrosion protection of zinc.

Table 1. Equivalent electrical circuit parameters of the zinc material, before and after corrosion.

Sample	Rs (Ω·cm²)	R1 (Ω·cm²)	R2 (Ω·cm²)	CPE1		CPE2	
				Y1 (Ω^{-1} cm^{-1} s^n)	n1	Y2 (Ω^{-1} cm^{-1} s^n)	n2
Bare Zinc	6.04 ± 0.17	102.2 ± 7.93	79.9 ± 24.32	0.000162 ± 3.47 × 10^{-5}	0.72 ± 0.015	0.01216 ± 0.008	0.77 ± 0.15
Hybrid-coated Zinc	12.11 ± 0.32	998 ± 29.89	899.5 ± 81.97	0.0000437 ± 2.15 × 10^{-6}	0.63 ± 0.005	0.00108 ± 0.000157	0.90 ± 0.061

The hybrid coating provides corrosion protection mainly by functioning as a compact and hydrophobic barrier. Alkoxy-silane and silanes undergo hydrolysis and subsequent condensation to form a highly crosslinked siloxane (Si–O–Si) network, which exhibits strong interfacial adhesion to the metallic substrate. In parallel, a methacrylic resin reinforces the organic phase of the coating, imparting additional toughness and flexibility that help mitigate cracking or mechanical failure. The synergistic contribution of silane groups and organic segments increases the hydrophobic character of the film, effectively reducing water uptake and restricting the permeation of aggressive species such as chloride ions and dissolved oxygen. These combined effects result in a robust and long-lasting protective layer that physically separates the metal surface from the surrounding corrosive environment [44–46].

The ionic conductivity of the coating was determined using impedance measurements. The ionic conductivity (δ) was calculated according to the following Equation:

$$\delta = \frac{L}{A \times R_b} \quad (1)$$

where L is the coating thickness, A is the exposed area, and R_b represents the coating resistance [47]. The ionic conductivity of the Hybrid-coated zinc in a 1 M NaCl solution was 0.013 mS/cm.

3.2. *Charge/Discharge Tests*

The effectiveness of the coating was then evaluated by long-term discharge tests performed in galvanostatic mode for a zinc-air battery. The zinc-air battery cell was composed of an air cathode, bare zinc or Hybrid-coated zinc as the anode and a separator, in NaCl solution. Figure 4 presents the discharge characteristics obtained with a Hybrid-coated and bare zinc electrode, at a discharge current density of 1 mA/cm^2, down to a threshold voltage of 0.5 V in a full-cell assembly. The obtained higher discharge capacity of the coated zinc electrode (43.7 mAh) compared to the bare zinc electrode (24.1 mAh) indicated that the coating successfully increased the corrosion resistance of the zinc anode, delaying its passivation process. By effectively slowing down passivation, the coating allows the zinc anode to maintain its active state for a longer period during discharge, resulting in higher capacity utilization. This is a pivotal factor in enhancing the performance and efficiency of zinc-air batteries, particularly for long-duration applications.

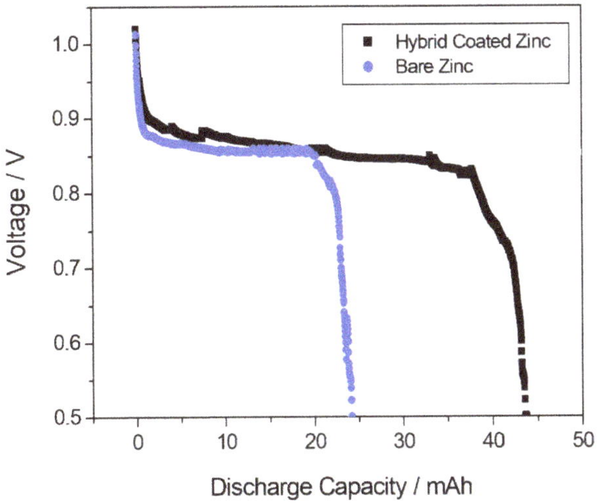

Figure 4. Discharge characteristics obtained with a Hybrid-coated and bare zinc electrode, at a discharge current density of 1 mA/cm^2.

The same self-fabricated laboratory cell was used to conduct several charge/discharge tests in 1 M NaCl electrolyte, in a symmetric cell configuration for assessing anode stability [48]. The charge/discharge behaviour of a cell consisting of Hybrid-coated or bare zinc vs. bare zinc was investigated at two current densities, 1 mA/cm^2 and 10 mA/cm^2.

The charge/discharge performances of zinc//zinc and coated zinc//zinc at the current density of 1 mA/cm^2 were examined and are presented in Figure 5. With the zinc//zinc cell the charge/discharge plots were very unstable with an overpotential, whereas Hybrid-coated Zn electrodes showed a stable cyclability. At a current density of 1 mA/cm^2, the symmetric cell with bare zinc electrodes exhibited a coulombic efficiency of 83.25% and a capacity retention of 88.68%, after 16 cycles, whereas the Hybrid-coated zinc versus bare zinc cell achieved values of 92.61% and 90.18%, respectively. These results indicate that the hybrid coating markedly improves the coulombic efficiency, also slightly improving capacity retention compared to bare zinc.

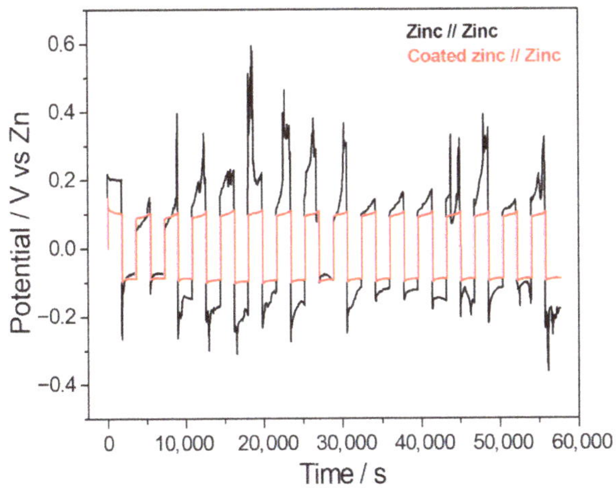

Figure 5. Charge/discharge behaviour of a cell composed of Hybrid-coated zinc//zinc and bare zinc//zinc measured at the current density of 1 mA/cm^2.

In Figure 6A, the charge/discharge behaviour of a cell composed of coated zinc vs. bare zinc at 1 mA/cm^2 and 10 mA/cm^2 is reported. The cell exhibited very stable anodic behaviour at 1 mA/cm^2. Stability in the anodic processes indicated that the electrode maintains its electrochemical activity without significant passivation or degradation [49–51]. At a current density of 10 mA/cm^2, the cell demonstrated a tendency towards passivation. In this case, the passivation phenomenon might hinder the desired electrochemical processes during the charge and discharge cycles. The coated zinc electrode appears to have limited effectiveness in preventing passivation or suppressing the formation of non-reactive layers on the electrode's surface. This suggests that the coating used on the zinc electrode may not be optimally designed to mitigate passivation effects at this particular current density. Passivation of Zn in neutral solutions proceeds via the gradual formation of $ZnO/Zn(OH)_2$ layers, which increases interfacial resistance [49–51]. In addition, the battery cell demonstrates very stable cathodic behaviour at both 1 and 10 mA/cm^2. Cathodic processes involve reduction reactions at the electrode. The cell exhibited nucleation and growth with a very limited tendency for outgrowth, which indicates highly stable cathodic behaviour. The coulombic efficiency and capacity retention of symmetric cells assembled with coated zinc vs. bare zinc electrodes were determined from galvanostatic charge–discharge measurements. At a current density of 10 mA/cm^2, the cell exhibited a coulombic efficiency of 91.07% and a capacity retention of 27.10%. When the current density was decreased to 1 mA/cm^2, the coulombic efficiency slightly improved to 92.61%, while the capacity retention markedly increased to 88.68%. This trend suggests that the coated zinc electrode exhibits enhanced stability and reversibility at lower current densities.

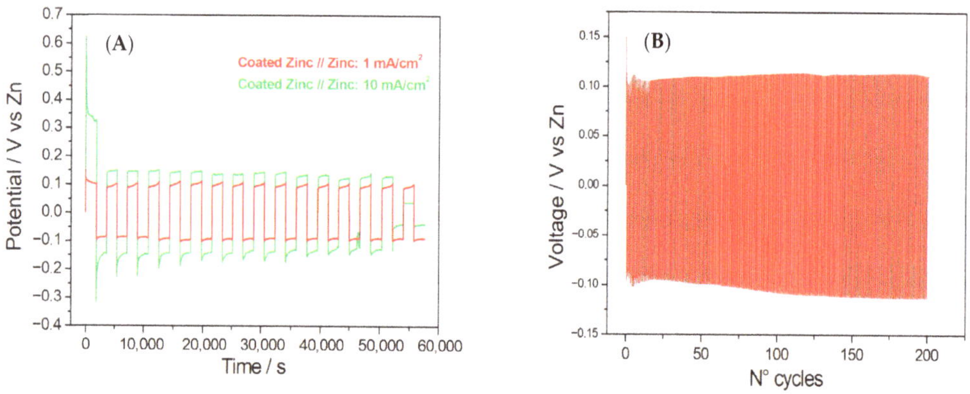

Figure 6. (**A**) Charge/discharge behaviour of a cell composed of Hybrid-coated zinc vs. bare zinc at 1 mA/cm^2 and 10 mA/cm^2; (**B**) Long-term galvanostatic charge/discharge cycles of Hybrid-coated zinc vs. bare zinc at 1 mA/cm^2 (1 h per cycle).

A long-term cyclability test with the coated zinc//zinc cell was also conducted at 1 mA/cm^2 (Figure 6B). A remarkable cycling stability is observed, sustaining 90.79% energy efficiency even after 200 cycles (1 h per cycle).

All in all, while the coating may not eliminate corrosion and passivation of the zinc electrode in secondary zinc-air batteries, its presence can significantly reduce their impact.

3.3. Morphological Studies

Figure 7 presents the 3D topographic images of bare and Hybrid-coated zinc samples obtained by AFM, before and after corrosion. The corroded surfaces correspond to those exposed to potentiostatic testing at −0.8 V for 60 min in 1 M NaCl solution. Before corrosion, both samples exhibited relatively smooth surfaces, with the coated sample displaying a noticeably smoother and more homogeneous morphology than the uncoated one.

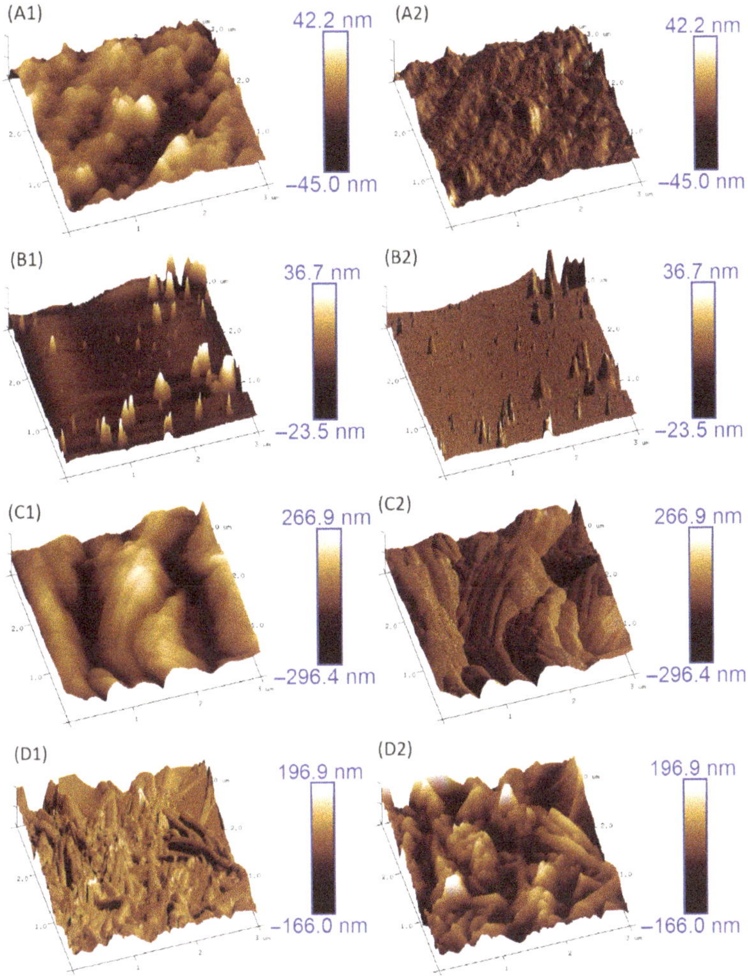

Figure 7. Height and Peak force Error AFM images of (**A1**,**A2**) Bare Zinc, (**B1**,**B2**) Hybrid-coated zinc, (**C1**,**C2**) Bare zinc after corrosion, and (**D1**,**D2**) Hybrid-Coated zinc after corrosion.

After the corrosion test, the differences in the surface condition between the coated and uncoated samples became more evident. The bare zinc surface was severely damaged due to corrosion, while the coated sample showed much less corrosion and some areas remained unaffected. This demonstrated the protective effect of the coating in mitigating corrosion and preserving the integrity of the underlying material by forming a barrier film on the surface.

The surface roughness (R_q) was evaluated with Nanoscope Analysis 1.5 software, calculated as the mean of ten distinct 500 nm × 500 nm regions within the imaged area. The results, along with the corresponding standard deviations, are summarized in Table 2. The obtained R_q values revealed the effectiveness of the coating in reducing surface roughness and preserving the smoothness of the coated sample, even after the corrosion test. The lower R_q values for the coated sample, relative to the uncoated one, further validate the protective and anticorrosive effectiveness of the applied coating.

Table 2. R_q values of zinc specimens without and with coating, before and after corrosion, extracted from AFM images reported in Figure 7.

Sample	R_q (nm)
Bare Zinc	2.99 ± 1.04
Hybrid-Coated Zinc	1.01 ± 0.37
Corroded Bare Zinc	32.52 ± 14.45
Corroded Hybrid-coated Zinc	2.28 ± 0.72

3.4. *Spectroscopic Studies*

The Raman spectroscopy analysis of the zinc samples, both coated and uncoated, before and after controlled corrosion obtained by potentiostatic tests carried out for 60 min at −0.8 V in 1 M NaCl solution, is shown in Figure 8. The Raman spectra provide valuable information on the chemical and structural changes that take place during the corrosive attack. The Raman spectra of the zinc sample before corrosion showed characteristic peaks associated with the formation of zinc oxide (ZnO) due to atmospheric exposure. The main observed band at 575 cm^{-1} could be assigned to the A_1(LO) mode, which is characteristic of zinc oxide. Additionally, there were peaks between 2850 cm^{-1} and 3000 cm^{-1} corresponding to symmetrical and asymmetrical $\nu(CH_3)$ vibrations. The spectrum of the coated zinc sample showed the same peaks related to zinc oxide formation, as well as additional peaks between 700 cm^{-1} and 1700 cm^{-1}, which are attributable to the organic polymer present in the coating [52]. Following the corrosive attack, both the coated and bare zinc samples showed clear peaks at 265 cm^{-1} and 400 cm^{-1}, as well as others at about 730 cm^{-1} and 900 cm^{-1}, which are indicative of the formation of $Zn(Cl)_2(Zn[OH]_2)_4$ (simonkolleite), a corrosion product [53,54]. The presence of simonkolleite was further confirmed by the peaks at 3460 cm^{-1} and 3495 cm^{-1}, which correspond to OH vibration modes. Additionally, the peak at 1078 cm^{-1} in the corroded zinc spectrum could be assigned to zinc carbonate [53]. Interestingly, after corrosion, the Raman spectrum of the coated zinc still displayed the characteristic bands of zinc oxide (ZnO), indicating that the coating provided some level of protection and slowed down the passivation process. However, in the uncoated zinc sample after corrosion, the oxide peaks were no longer present, indicating that the corrosion had progressed more extensively in the absence of protection.

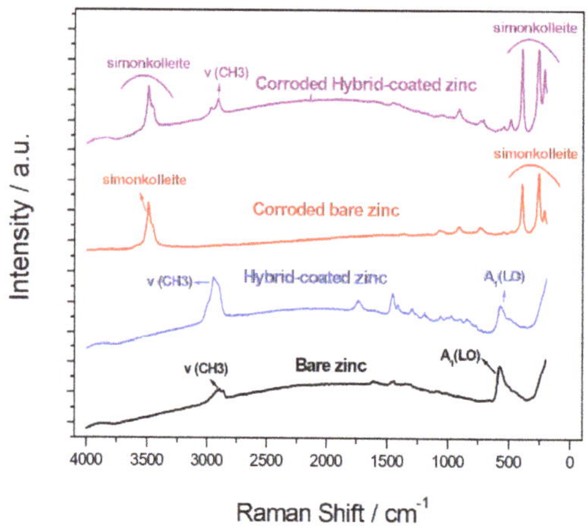

Figure 8. Raman spectra of the bare and coated samples before and after corrosion.

Furthermore, the peaks associated with the organic polymer coating were still present in the coated zinc sample after corrosion, although they were less evident. These results suggested that the protective hybrid coating on the zinc electrode provides some level of resistance against corrosive attack, slowing down passivation without blocking corrosion through a barrier mechanism [44–46]. This is a positive aspect of recharging the zinc electrode in the zinc-air battery, as it allows for a more controlled and stable electrochemical performance.

3.5. XRD Studies

An XRD analysis was performed on the Hybrid-coated zinc sample after the potentiostatic corrosion test at −0.8 V for 1 h, and the corresponding diffraction pattern is shown in Figure 9, together with that of the coating alone. The characteristic zinc diffraction peaks at 36.4°, 38.9°, 43.2°, 54.5°, 70.5°, and 77.6° correspond to the International Centre for Diffraction Data (JCPDS card No. 03-065-5973) and remain visible even after the corrosion test. The broad amorphous signal centered around 23° is attributed to the coating. Its persistence on the Hybrid-coated zinc after corrosion confirms the protective effectiveness of the Hybrid coating.

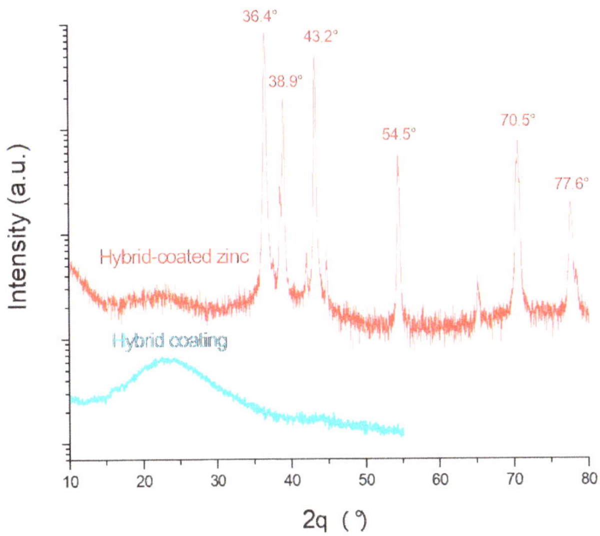

Figure 9. XRD spectra of the bare and coated samples after corrosion.

4. Conclusions

In this study, a hybrid coating was applied on the surface of a zinc electrode to enhance the corrosion resistance in a neutral electrolyte. The effectiveness of this approach was investigated through electrochemical and spectroelectrochemical techniques.

The electrochemical results indicated that the applied coating was successful in reducing the corrosion current density, thus enhancing the corrosion resistance of the zinc electrode. Morphological studies and roughness measurements using Atomic Force Microscopy further supported the reduced corrosion susceptibility of the coated zinc sample. Raman spectroscopy combined with XRD studies demonstrated that the coating slowed down the passivation process without blocking corrosion, and the hybrid coating provides corrosion protection mainly by functioning as a compact and hydrophobic barrier.

Long-term charge/discharge tests of the coated zinc vs. zinc at the current density of 1 mA/cm^2 showed a remarkable cycle stability even after 200 cycles with a coulombic efficiency of 90.79%. This allowed for a greater discharge capacity, making the coating advantageous for the recharging of the zinc electrode in the zinc-air battery.

Author Contributions: Conceptualization, S.B., B.B., C.E.C., R.S. and C.M.; methodology, S.B., B.B., C.E.C., R.S. and C.M.; validation, S.B. and C.M.; investigation, S.B., B.B., C.E.C., R.S. and C.M.; data curation, S.B., C.E.C., R.S. and C.M.; writing—original draft preparation, S.B. and C.M.; writing—review and editing, S.B. and C.M.; supervision, B.B., C.E.C. and C.M. All authors have read and agreed to the published version of the manuscript.

Funding: This research is supported by PRIN2022 PNRR "BAT-MEN" (BATtery Modeling, Experiments & Numerics) Enhancing battery lifetime, Project code: P20228C2PP 001, CUP: F53D23010020001, funded by MIUR (Italian Ministry of University and Research) and European Union–NextGenerationEU.

Data Availability Statement: The original contributions presented in this study are included in the article. Further inquiries can be directed to the corresponding authors.

Conflicts of Interest: The authors declare no conflicts of interest.

References

1. Yuan, Z.; Yin, Y.; Xie, C.; Zhang, H.; Yao, Y.; Li, X. Advanced Materials for Zinc-Based Flow Battery: Development and Challenge. *Adv. Mater.* **2019**, *31*, 1902025. [CrossRef] [PubMed]
2. Lu, W.; Xie, C.; Zhang, H.; Li, X. Zinc dendrites Inhibition for Zinc-based Battery. *ChemSusChem* **2018**, *11*, 1–12. [CrossRef] [PubMed]
3. Wang, C.; Yu, Y.; Niu, J.; Liu, Y.; Bridges, D.; Liu, X.; Pooran, J.; Zhang, Y.; Hu, A. Recent Progress of Metal—Air Batteries—A Mini Review. *Appl. Sci.* **2019**, *9*, 2787. [CrossRef]
4. Mele, C.; Bozzini, B. Spectroelectrochemical investigation of the anodic and cathodic behaviour of zinc in 5.3 M KOH. *J. Appl. Electrochem.* **2015**, *45*, 43–50. [CrossRef]
5. Mele, C.; Bilotta, A.; Bocchetta, P.; Bozzini, B. Characterization of the particulate anode of a laboratory flow Zn–air fuel cell. *J. Appl. Electrochem.* **2017**, *47*, 877–888. [CrossRef]
6. Bozzini, B.; Kazemian, M.; Kiskinova, M.; Kourousias, G.; Mele, C.; Gianoncelli, A. Operando soft X-ray microscope study of rechargeable Zn–air battery anodes in deep eutectic solvent electrolyte. *X-Ray Spectrom.* **2019**, *48*, 527–535. [CrossRef]
7. Mu, T.; Lu, H.; Wang, H.; Wang, Y.; Lou, S.; Zhang, Y.; Gao, J.; Ma, Y.; Zuo, P.; Zhu, X.; et al. Artificial organic-inorganic hybrid interface enables reversible Zn anodes. *Nano Energy* **2025**, *138*, 110835. [CrossRef]
8. Suppanucroa, N.; Yoopensuk, W.; Pimoei, J.; Thanapong-a-morn, W.; Kao-Ian, W.; Pakawanit, P.; Mahlendorf, F.; Kheawhom, S.; Somwangthanaroj, A. Enhanced long-term stability of zinc-air batteries using a quaternized PVA-chitosan composite separator with thin-layered MoS_2. *Electrochim. Acta* **2025**, *510*, 145361. [CrossRef]
9. Shinde, S.S.; Kim, S.H.; Wagh, N.K.; Lee, J.H. Design Strategies for Practical Zinc-Air Batteries Toward Electric Vehicles and beyond. *Adv. Energy Mater.* **2025**, *15*, 2405326. [CrossRef]
10. Wang, Y.; Kwok, H.; Pan, W.; Zhang, H.; Leung, D.Y.C. Innovative paper-based Al-air batteries as a low-cost and green energy technology for the miniwatt market. *J. Power Sources* **2019**, *414*, 278–282. [CrossRef]
11. Mainar, A.R.; Leonet, O.; Bengoechea, M.; Boyano, I.; de Meatza, I.; Kvasha, A.; Guerfi, A.; Blázquez, J.A. Alkaline aqueous electrolytes for secondary zinc-air batteries: An overview. *Int. J. Energy Res.* **2016**, *40*, 1032–1049. [CrossRef]
12. Linden, D.; Reddy, T.B. *Handbook of Batteries*, 3rd ed.; TRD McGraw-Hill: Singapore, 2002.
13. Bozzini, B.; Altissimo, M.; Amati, M.; Bocchetta, P.; Gianoncelli, A.; Gregoratti, L.; Kourousias, G.; Mancini, L.; Mele, C.; Kiskinova, M. In situ and ex situ X-ray microspectroelectrochemical methods for the study of zinc-air batteries. *Encycl. Interfacial Chem. Surf. Sci. Electrochem.* **2018**, 174–194. [CrossRef]
14. Bozzini, B.; Mele, C.; Veneziano, A.; Sodini, N.; Lanzafame, G.; Taurino, A.; Mancini, L. Morphological Evolution of Zn-Sponge Electrodes Monitored by in Situ X-ray Computed Microtomography. *ACS Appl. Energy Mater.* **2020**, *3*, 4931–4940. [CrossRef]
15. Zhao, H. Recent Advances in Rechargeable Zn-Air Batteries. *Molecules* **2024**, *29*, 5313. [CrossRef]
16. Durmus, Y.E.; Guerrero, S.S.M.; Tempel, H.; Hausen, F.; Kungl, H.; Eichel, R.A. Influence of Al Alloying on the Electrochemical Behavior of Zn Electrodes for Zn–Air Batteries with Neutral Sodium Chloride Electrolyte. *Front. Chem.* **2019**, *7*, 800. [CrossRef]
17. Guerrero, S.S.M.; Durmus, Y.E.; Dzieciol, K.; Basak, S.; Tempel, H.; van Waasen, S.; Kungl, H.; Eichel, R.A. Improved Electrochemical Performance of Zinc Anodes by EDTA in Near-Neutral Zinc−Air Batteries. *Batter. Supercaps* **2021**, *4*, 1830–1842. [CrossRef]

18. Wei, W.; Xu, J.; Chen, W.; Mi, L.; Zhang, J. A review of sodium chloride-based electrolytes and materials for electrochemical energy technology. *J. Mater. Chem. A* **2022**, *10*, 2637–2671. [CrossRef]
19. Jindra, J.; Mrha, J.; Musilová, M. Zinc-air cell with neutral electrolyte. *J. Appl. Electrochem.* **1973**, *3*, 297–301. [CrossRef]
20. Amendola, S.; Binder, M.; Black, P.J.; Sharp-Goldman, S.; Johnson, L.; Kunz, M.; Oster, M.; Chciuk, T.; Johnson, R. Electrically Rechargeable, Metal-Air Battery Systems and Methods. US Patent US20120021303A1, 26 January 2012.
21. Goh, F.W.T.; Liu, Z.; Hor, T.S.A.; Zhang, J.; Ge, X.; Zong, Y.; Yu, A.; Khoo, W. A Near-Neutral Chloride Electrolyte for Electrically Rechargeable Zinc-Air Batteries. *J. Electrochem. Soc.* **2014**, *161*, A2080–A2086. [CrossRef]
22. Sumboja, A.; Ge, X.; Zheng, G.; Goh, F.W.T.; Hor, T.S.A.; Zong, Y.; Liu, Z. Durable rechargeable zinc-air batteries with neutral electrolyte and manganese oxide catalyst. *J. Power Sources* **2016**, *332*, 330–336. [CrossRef]
23. Clark, S.; Latz, A.; Horstmann, B. Rational Development of Neutral Aqueous Electrolytes for Zinc–Air Batteries. *ChemSusChem* **2017**, *10*, 4735–4747. [CrossRef]
24. Yu, J.; Zhao, C.-X.; Liu, J.-N.; Li, B.-Q.; Tang, C.; Zhang, Q. Seawater-based electrolyte for zinc–air batteries. *Green. Chem. Eng.* **2020**, *1*, 117–123. [CrossRef]
25. Neburchilov, V.; Wang, H.; Martin, J.J.; Qu, W. A review on air cathodes for zinc-air fuel cells. *J. Power Sources* **2010**, *195*, 1271–1291. [CrossRef]
26. Gu, P.; Zheng, M.; Zhao, Q.; Xiao, X.; Xue, H.; Pang, H. Rechargeable zinc-air batteries: A promising way to green energy. *J. Mater. Chem. A* **2017**, *5*, 7651–7666. [CrossRef]
27. Cao, R.; Lee, J.S.; Liu, M.; Cho, J. Recent progress in non-precious catalysts for metal-air batteries. *Adv. Energy Mater.* **2012**, *2*, 816–829. [CrossRef]
28. Xue, G.; Bian, H.; Wang, B.; Wu, H.; Li, F.; Wang, C.; Zhou, Q.; Jia, S.; Hu, Z.; Ma, Y.; et al. An organic-inorganic composite coating with excellent protection and ion regulation effects for highly stable zinc anodes. *Colloids Surf. A Physicochem. Eng. Asp.* **2024**, *703*, 135272. [CrossRef]
29. Li, B.; Yang, H.; He, J.; Yu, S.; Xiao, R.; Luo, H.; Wen, Y.; Peng, S.; Liao, X.; Yang, D. Photopolymerization of Coating Materials for Protection against Carbon Steel Corrosion. *Materials* **2023**, *16*, 2015. [CrossRef]
30. Corcione, C.E.; Striani, R.; Frigione, M. Organic-inorganic UV-cured methacrylic-based hybrids as protective coatings for different substrates. *Prog. Org. Coat.* **2014**, *77*, 1117–1125. [CrossRef]
31. Striani, R.; Cappai, M.; Casnedi, L.; Corcione, C.E.; Pia, G. Coating's influence on wind erosion of porous stones used in the Cultural Heritage of Southern Italy: Surface characterisation and resistance. *Case Stud. Constr. Mater.* **2022**, *17*, e01501. [CrossRef]
32. Pia, G.; Corcione, C.E.; Striani, R.; Casnedi, L.; Sanna, U. Coating's influence on water vapour permeability of porous stones typically used in cultural heritage of Mediterranean area: Experimental tests and model controlling procedure. *Prog. Org. Coat.* **2017**, *102*, 239–246. [CrossRef]
33. Striani, R.; Corcione, C.E.; Muia, G.D.; Frigione, M. Durability of a sunlight-curable organic–inorganic hybrid protective coating for porous stones in natural and artificial weathering conditions. *Prog. Org. Coat.* **2016**, *101*, 1–14. [CrossRef]
34. Fouassier, J.P. An introduction to the basic principles in UV-curing. *Radiat. Curing Polym. Sci. Technol.* **1993**, *1*, 49–113.
35. Corcione, C.E.; Frigione, M.; Striani, R. Hybrid Organic-Inorganic Nanostructured UV-Curable Formulation and Method for Preparation Thereof. European Patent 13001868.2, 11 April 2013.
36. Bozzini, B.; Bagheri, S.; Boniardi, M.; Mancini, L.; Marini, E.; Sgura, I.; Mele, C. Quantifying and rationalizing polarization curves of Zn-air fuel-cells: A simple enabling contribution to device-scale analysis and monitoring. *Electrochim. Acta* **2022**, *425*, 140712. [CrossRef]
37. Mele, C.; Bozzini, B. Corrosion performance of austenitic stainless steel bipolar plates for Nafion- and room-temperature ionic-liquid-based PEMFCs. *Open Fuels Energy Sci. J.* **2012**, *5*, 47–52. [CrossRef]
38. Pakshir, M.; Bagheri, T.; Kazemi, M.R. In vitro evaluation of the electrochemical behaviour of stainless steel and Ni-Ti orthodontic archwires at different temperatures. *Eur. J. Orthod.* **2013**, *35*, 407–413. [CrossRef] [PubMed]
39. Meng, Y.; Liu, L.; Zhang, D.; Dong, C.; Yan, Y.; Volinsky, A.A.; Wang, L.N. Initial formation of corrosion products on pure zinc in saline solution. *Bioact. Mater.* **2019**, *4*, 87–96. [CrossRef]
40. Ahaliabadeh, Z.; Miikkulainen, V.; Mäntymäki, M.; Mousavihashemi, S.; Lahtinen, J.; Lide, Y.; Jiang, H.; Mizohata, K.; Kankaanpää, T.; Kallio, T. Understanding the Stabilizing Effects of Nanoscale Metal Oxide and Li-Metal Oxide Coatings on Lithium-Ion Battery Positive Electrode Materials. *ACS Appl. Mater. Interfaces* **2021**, *13*, 42773–42790. [CrossRef]
41. Duchoslav, J.; Steinberger, R.; Arndt, M.; Keppert, T.; Luckeneder, G.; Stellnberger, K.H.; Hagler, J.; Angeli, G.; Riener, C.K.; Stifter, D. Evolution of the surface chemistry of hot dip galvanized Zn-Mg-Al and Zn coatings on steel during short term exposure to sodium chloride containing environments. *Corros. Sci.* **2015**, *91*, 311–320. [CrossRef]
42. Liu, L.; Li, P.; Zou, Y.; Luo, K.; Zhang, F.; Zeng, R.C.; Li, S. In vitro corrosion and antibacterial performance of polysiloxane and poly(acrylic acid)/gentamicin sulfate composite coatings on AZ31 alloy. *Surf. Coat. Technol.* **2016**, *291*, 7–14. [CrossRef]
43. Liu, J.C.; Park, S.W.; Nagao, S.; Nogi, M.; Koga, H.; Ma, J.S.; Zhang, G.; Suganuma, K. The role of Zn precipitates and Cl- anions in pitting corrosion of Sn-Zn solder alloys. *Corros. Sci.* **2015**, *92*, 263–271. [CrossRef]

44. Longhi, M.; Kunsta, S.R.; Beltrami, L.V.R.; Kerstner, E.K.; Filho, C.I.S.; Sarmento, V.H.V.; Malfatti, C. Effect of tetraethoxy-silane (TEOS) amounts on the corrosion prevention properties of siloxane-pmma hybrid coatings on galvanized steel substrates. *Mater. Res.* **2015**, *18*, 1140–1155. [CrossRef]
45. Figueira, R.B.; Silva, C.J.R.; Pereira, E.V. Organic–inorganic hybrid sol–gel coatings for metal corrosion protection: A review of recent progress. *J. Coat. Technol. Res.* **2015**, *12*, 1–35. [CrossRef]
46. Harb, S.V.; Trentin, A.; Torrico, R.F.O.; Pulcinelli, S.H.; Santilli, C.V.; Hammer, P. Organic-Inorganic Hybrid Coatings for Corrosion Protection of Metallic Surfaces. *New Technol. Prot. Coat.* **2017**, *290*, 153–162. [CrossRef]
47. Fan, X.; Liu, J.; Song, Z.; Han, X.; Deng, Y.; Zhong, C.; Hu, W. Porous nanocomposite gel polymer electrolyte with high ionic conductivity and superior electrolyte retention capability for long-cycle-life flexible zinc–air batteries. *Nano Energy* **2019**, *56*, 454–462. [CrossRef]
48. Li, Q.; Chen, A.; Wang, D.; Pei, Z.; Zhi, C. "Soft Shorts" Hidden in Zinc Metal Anode Research. *Joule* **2022**, *6*, 273–279. [CrossRef]
49. Fuchs, D.; Müller, C.; Schaffeld, M.; Mahlendorf, F.; Hoster, H.E. New Insights Into Zinc Passivation Through Operando Measured Zincate Concentrations. *Batter. Supercaps* **2024**, *7*, e202400298. [CrossRef]
50. Bockelmann, M.; Reining, L.; Kunz, U.; Turek, T. Electrochemical characterization and mathematical modeling of zinc passivation in alkaline solutions: A review. *Electrochim. Acta* **2017**, *237*, 276–298. [CrossRef]
51. Zhang, Y.; Zheng, X.; Wang, N.; Lai, W.H.; Liu, Y.; Chou, S.L.; Liu, H.K.; Dou, S.X.; Wang, Y.X. Anode optimization strategies for aqueous zinc-ion batteries. *Chem. Sci.* **2022**, *13*, 14246–14263. [CrossRef]
52. Nikitenko, V.A.; Plekhanov, V.G.; Mukhin, S.V. Raman Spectra of Oxide zinc powder and single crystals. *J. Appl. Spectrosc.* **1996**, *63*, 350–352. [CrossRef]
53. Gu, R.A.; Shen, X.Y.; Liu, G.K.; Ren, B.; Tian, Z.Q. Surface-enhanced raman scattering from bare Zn electrode. *J. Phys. Chem. B* **2004**, *108*, 17519–17522. [CrossRef]
54. Cai, D.; Neyer, A.; Kuckuk, R.; Heise, H.M. Raman, mid-infrared, near-infrared and ultraviolet-visible spectroscopy of PDMS silicone rubber for characterization of polymer optical waveguide materials. *J. Mol. Struct.* **2010**, *976*, 274–281. [CrossRef]

MDPI AG
Grosspeteranlage 5
4052 Basel
Switzerland
Tel.: +41 61 683 77 34

Energies Editorial Office
E-mail: energies@mdpi.com
www.mdpi.com/journal/energies

www.ingramcontent.com/pod-product-compliance
Lightning Source LLC
Chambersburg PA
CBHW062038220926
51850CB00010B/2598